89 Topics in Current Chemistry
Fortschritte der Chemischen Forschung

AF840158

Plasma Chemistry I

Editors: S. Veprek and M. Venugopalan

Springer-Verlag
Berlin Heidelberg GmbH 1980

This series presents critical reviews of the present position and future trends in modern chemical research. It is addressed to all research and industrial chemists who wish to keep abreast of advances in their subject.

As a rule, contributions are specially commissioned. The editors and publishers will, however, always be pleased to receive suggestions and supplementary information. Papers are accepted for "Topics in Current Chemistry" in English.

ISBN 978-3-662-15976-7 ISBN 978-3-540-38974-3 (eBook)
DOI 10.1007/978-3-540-38974-3

Library of Congress Cataloging in Publication Data. Main entry under title: Plasma chemistry. (Topics in current chemistry ; 89–90) Bibliography: v. 1, p. ; v. 2, p. Includes indexes. 1. Plasma chemistry – Addresses, essays, lectures. I. Series. QD1.F58 vol. 89–90 [QD581] 540'.8s [541'042'4] 79–25770

Originally published by Springer-Verlag Berlin Heidelberg New York in 1980
Softcover reprint of the hardcover 1st edition 1980

Typesetting and printing: Schwetzinger Verlagsdruckerei GmbH, 6830 Schwetzingen. Bookbinding: Konrad Triltsch, Graphischer Betrieb, 8700 Würzburg
2152/3140 – 543210

Contents

Springer-Verlag, Postfach 105280 · D-6900 Heidelberg 1, Telephone (06221) 487-1 · Telex 04-61723

Heidelberger Platz 3 · D-1000 Berlin 33, Telephone (030) 822001 · Telex 01-83319

Springer-Verlag, New York Inc., 175, Fifth Avenue · New York, NY 10010, Telephone 477-8200

Preface

During the past ten years significant developments have taken place in the field of plasma chemistry. Progress of research, or the lack of it in certain areas, was discussed at a number of Gordon conferences, and at four international symposia, several round tables and topical meetings sponsored by IUPAC. The editors of *Topics in Current Chemistry* have therefore undertaken to bring the advances into perspective by issuing volumes which are subtitled *Plasma Chemistry*. The volumes will deal with topics which are the main concern of many researchers. Such topics include elementary processes in low pressure plasmas, plasma-material interactions, plasma polymerization and thermal plasma processing. Because of the ever increasing importance of energy and environmental problems the editors have found it only fit to begin the series with a chapter on elementary plasma reactions of environmental interest.

Among the significant developments mention may be made of the realization that electron-impact induced excitation of vibrational states and subsequent v-v climbing is the most important dissociation process in many gases under conditions obtained in low pressure plasmas. A successful recent application of the low pressure plasma is in the preparation of materials with extreme requirements of purity, homogeneity and radial profile of refractive index, all of which are necessary in the manufacture of optical wave guides. Plasma etching and polymerization technologies are rapidly emerging and elementary processes occuring at solid surfaces immersed in plasmas are receiving considerable attention. Investigations of plasma-material interactions have contributed to a better understanding of the impurity release and have led to suggestions for future control of impurities in magnetically confined thermonuclear machines for nuclear fusion research. Thermal plasmas have been applied in ceramics, powder technology and extractive metallurgy; in the production of solar grade silicon, in nitric oxide synthesis, and in the gasification and desulfurization of coal.

It is hoped that the volumes will appeal to a broad spectrum of readers and will promote further research in the field of plasma chemistry.

The editors acknowledge the help and cooperation of Dr. F. L. Boschke and his staff at Springer-Verlag. They are grateful to the various authors for their timely contributions and to their colleagues and families for encouragement during the preparation of the volumes.

S. Vepřek
M. Venugopalan

Elementary Plasma Reactions of Environmental Interest

David Smith and Nigel G. Adams

Department of Space Research, University of Birmingham, Birmingham, B 15 2TT, England

Table of Contents

1 Introduction

The atmosphere, which extends to many thousands of kilometres from the Earth's surface before it merges with the very tenuous interplanetary gas, is immersed in a radiation field mainly generated by the sun but which includes galactic cosmic ray components. Energetic short wavelength components of the radiation field ensure that the atmosphere is at least partially ionized at all times, the degree of ionization varying by many orders of magnitude throughout the atmosphere. The production of very reactive ions, electrons and neutral radicals initiates an extremely complex gas-phase chemistry which can modify greatly the composition of the atmosphere, thus generating the perceived atmospheric environment.

In the upper reaches of the atmosphere, more specifically in the magnetosphere and upper ionosphere, the degree of ionization is relatively high, the charged particles are electrons and positive ions only and the gas pressure is very low. The plasma properties of these regions are very evident and they have received considerable attention from physicists over the last few decades principally those concerned with radiowave propagation[1)]. The chemistry of these regions is relatively simple and reasonably well understood, being described by just a few binary reaction processes (Sect. 3.2.2 and 4). Descending through the atmosphere into the middle and lower ionosphere, through the stratosphere and troposphere, the gas pressure increases approximately exponentially with decreasing altitude. Negative ions ultimately replace electrons as the negatively charged species and thus a positive ion/negative ion plasma exists in which the charged particle/neutral particle ratio is very small because of the high pressure which obtains but in which the absolute charged particle density still is appreciable (Sect. 2.2). Thus the chemistry becomes much more complex. Ternary reactions, the rate coefficients of which are known to be very sensitive to temperature, become very important. Also the presence of a large number of different neutral molecular species variously generated (biogenetically, man-made pollutants, etc.) which are efficiently photolysed by solar radiation, ensure that a rich ion and a parallel neutral chemistry will be continuously occurring, the latter dominating the chemical evolution in the stratosphere and the troposphere. Nevertheless, the presence of significant concentrations of charged particles coupled with their greater reactivity will have a significant influence on chemical evolution in the medium. It is with the elementary charged particle reactions which are occurring continuously in the environmental plasma that this review is concerned.

Progress towards an understanding of ionospheric ion chemistry has resulted from in-situ determinations of ion composition using balloon-, rocket- and satellite-borne mass spectrometers coupled with the development and exploitation of laboratory experiments to determine reaction rate coefficients, ion products, etc. Progress has been much slower with respect to stratospheric and tropospheric chemistry, not least due to the difficulties inherent in mass spectrometer sampling of these relatively high pressure regions. However, significant advances have been made recently both in in-situ sampling techniques and in the acquisition of relevant laboratory data.

In order that the discussion of the individual reaction processes, which combined to describe the overall chemistry of the atmospheric regions, can be facilitated and readily placed into context, a brief description of the atmospheric plasma medium (Sect. 2) will be presented, including the neutral composition, the ionizing radiations and the ion composition in the various regions. Then, following the detailed discussion of elementary processes in Sect. 3, which will concentrate on the most recent developments, the paper will be concluded (Sect. 4) by a brief summary of the current status of atmospheric ion chemistry.

2 Description of the Environmental Plasma

The combined influence of gravity and solar radiations on the Earth's atmosphere results in several distinct 'regions' which are designated according to such physical parameters as the ambient temperature and the ionization density. Photoionization and a subsequent complicated ion chemistry converts the simple primary positive and negative ions into the observed, generally more complex ions.

2.1 The Neutral Atmosphere: Composition and Temperature

Much has been written concerning the structures and properties of the neutral atmosphere [2–5], but the details are not relevant here. Only those general features which significantly influence the chemistry of the atmosphere will be discussed and then only very briefly.

In the absence of temperature gradients, that is for an isothermal atmosphere, the total pressure or total gas density would decrease very nearly exponentially with increasing altitude. However radiant heating ensures that such a simple structure does not pertain in the Earth's atmosphere and the total gas density decreases in the manner shown in Fig. 1. The altitudinal variation of ambient temperature is even more complicated and in the upper reaches of the atmosphere is a result of the absorption of short wavelength radiation whilst in the lower atmosphere it is due to the heating of the Earth's surface by longer wavelength radiation followed by conduction and convection. A typical neutral temperature versus altitude profile is shown in Fig. 2. From sea level, the *atmospheric temperature*, T_g, at first decreases with altitude as the distance from the heat source (the Earth's surface) increases. At about 15 km altitude a minimum in T_g appears, the so-called tropopause, the actual altitude of the tropopause being subject to both diurnal and seasonal variations as are most of the atmospheric physical features.

The increasing T_g at altitudes above the tropopause is largely due to the photochemical production of ozone via the well-known reactions [7]

$$O_2 \xrightarrow{h\nu} O + O \tag{1a}$$

$$O + O_2 + M \longrightarrow O_3 + M \tag{1b}$$

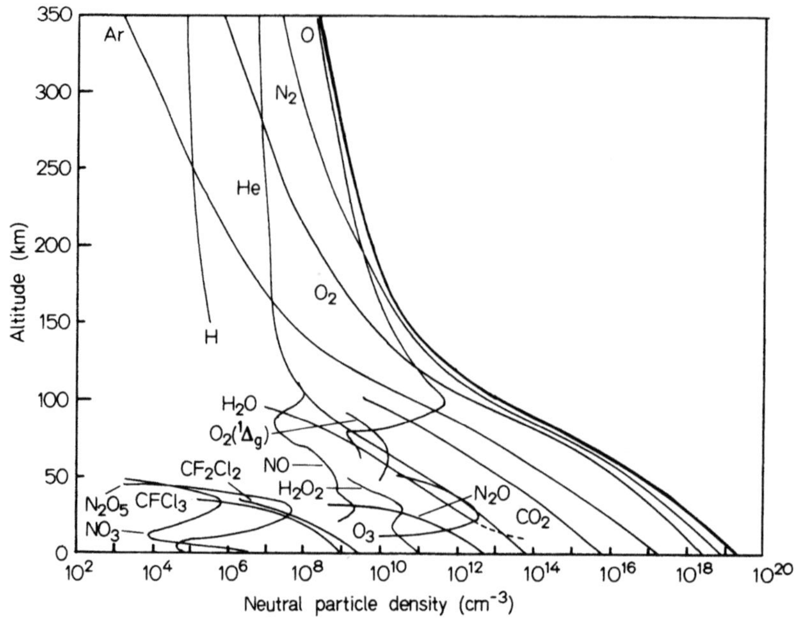

Fig. 1. The neutral composition of the Earth's atmosphere as a function of altitude. The sources of the altitudinal profile data are: H, He, O, N_2, O_2, Ar, Ref.[6]; O_3, Ref.[7]; H_2O, Refs.[8] and [10]; NO, Refs.[9] and [10]; $O_2(^1\Delta_g)$, Ref.[11]; O below 90 km, Ref.[12]; CF_2Cl_2 and $CFCl_3$, Ref.[13]; N_2O, Ref.[14]; H_2O_2, NO_3 and N_2O_5, from model calculations given in Ref.[15]; CO_2 is assumed to have a constant mixing ratio of 300 p.p.m. Clearly many other minority species are present in the lower atmosphere (e.g. OH, HO_2; see Ref.[15]) but these have been omitted for clarity. The thick line indicates the total neutral particle density

M can be any third body which carries away the energy liberated in the reaction but almost invariably will be N_2 or O_2. Destruction of the ozone also results in atmospheric heating:

$$O_3 \xrightarrow{h\nu} O_2 + O \quad (2a)$$

$$O_3 + O \longrightarrow O_2 + O_2 \quad (2b)$$

Reactions (1) and (2) essentially convert solar radiant energy into thermal energy. The parameters which determine the rate of ozone formation (UV photon flux, atomic and molecular oxygen number density and the total gas number density) are not constant with altitude and so the ozone concentration and hence T_g varies with altitude. The net result is that T_g increases thoughout the stratosphere until a maximum is reached at the stratopause whence T_g begins to decrease again.

The presence of the ozone layer ensures that solar UV radiation (in the wavelength range 2–300 nm) which is harmful to living things is prevented [by reaction (2a)] from reaching the Earth's surface. Since ozone is known from laboratory studies to be destroyed by some atmospheric pollutants and the photofragments thereof[17], pollution of the atmosphere (particularly by freons) has recently been the

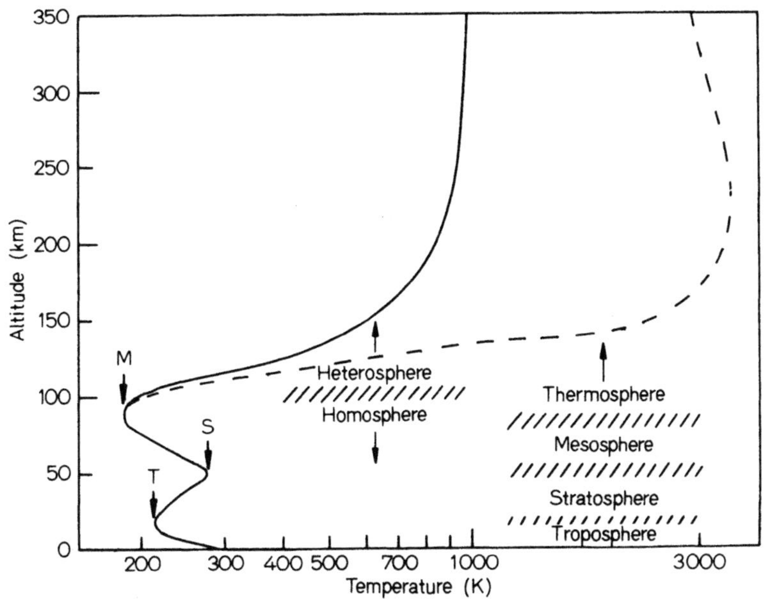

Fig. 2. A typical altitudinal variation of the neutral gas temperature[6] (*solid curve*) and the plasma electron temperature[16] (*dashed curve*) in the Earth's atmosphere. Note that the temperature scale (*absissa*) is logarithmic. The atmospheric regions are as indicated and M, S and T indicate the mesopause, stratopause and tropopause, respectively

cause of much discussion. This has led to experiments specifically designed to determine the concentration of such pollutants in the troposphere and the stratosphere [18]. The various techniques which have been used for this and some of the results obtained have recently been reviewed by Ehhalt[13].

The region of the atmosphere above the stratopause in which T_g continuously reduces towards a second minima, the mesopause, is termed the *mesosphere.* The mesopause occurs at an altitude of about 80 km and is the lowest temperature point in the atmosphere, values of T_g approaching 100 K having been recorded, although temperatures nearer to 200 K are more typical. Significantly, it is the relatively high pressure, low temperature mesosphere in which "cluster" ions were first detected in the atmosphere (Sect. 2.2).

Above the mesopause, T_g increases rapidly. In this region, termed the *thermosphere* (Fig. 2), absorption of short wavelength solar radiation is occurring (Fig. 3) which results in the efficient photodissociation of molecular oxygen, and the photoionization of the O atoms so produced and of the O_2 and N_2 molecules. Thus, T_g increases beyond 1000 K, approaching 2000 K at times. Whereas below 100 km the neutral gas particles, the ions and the electrons in the plasma all possess the same kinetic temperature, above 100 km, due to the lower pressure and the subsequent reduced electron/heavy particle collision frequency and the large amount of energy imparted to the photoelectrons, the electron temperature, T_e increases above T_g (and T_i the ion temperature, which is $\sim T_g$, see Fig. 2).

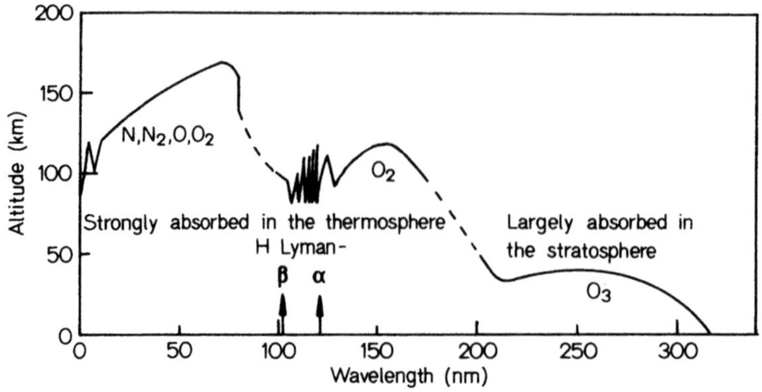

Fig. 3. The absorption spectrum for solar radiation in the Earth's atmosphere[19]. On the ordinate is plotted the altitude at which the radiation intensity is reduced by a factor e^{-1} from its unattenuated value. The species predominantly responsible for the absorption in the various wavelength ranges are as indicated. The wavelength of the H Lyman-α radiation closely coincides with a window in the O_2 absorption spectrum

The temperature gradients, gravity and photochemistry exercise an important influence on the neutral composition of the atmosphere. The mesopause divides the atmosphere into two distinct regions in terms of their gas dynamical properties. Below about 100 km, large scale dynamical effects result in mass transport throughout the region, i. e. effective mixing of the major and the more stable minor neutral constituents (e. g., CO_2, H_2O) occurs and the "mixing ratio" of a particular species remains approximately constant. Hence the term "homosphere" is often applied to this region. The transport of minor neutral constituents from the troposphere into the stratosphere and mesosphere has a profound influence on the chemistry of these regions.

Above an altitude of about 100 km (in the thermosphere), convective mixing of the various gases of differing molecular weight is no longer effective and gravitational separation of the neutral species becomes evident (Fig. 1). Thus the relative concentrations of the lighter species H, He and O increases with increasing altitude. Hence the term *"heterosphere"* is often applied to this region of the atmosphere. No complex molecules can survive in the heterosphere because of the prevailing intense solar radiation and so the chemistry is relatively simple (Sect. 3.2.2 and 4).

In summary, large variations in the pressure, composition and temperature exist in the atmosphere, the greatest variations occurring in the lower atmospheric regions which, as will be seen, are reflected by the much more complex chemistry occurring in these regions, a chemistry which is thought to be dominated by the minor neutral constituents, doubtlessly some of which have yet to be identified.

2.2 The Ionized Atmosphere: Ion Composition and Number Density

The presence of ionization in the upper atmosphere was first demonstrated about half a century ago by reflecting radio waves from the ambient plasma [1], and by this technique it was shown that the ionosphere was clearly structured, several "ledges" in the electron density being evident (Fig. 4). Below 80 km, no radio reflection could be detected and it was concluded that no significant ionization existed below that altitude. This conclusion is now known to be incorrect since, although the ionization

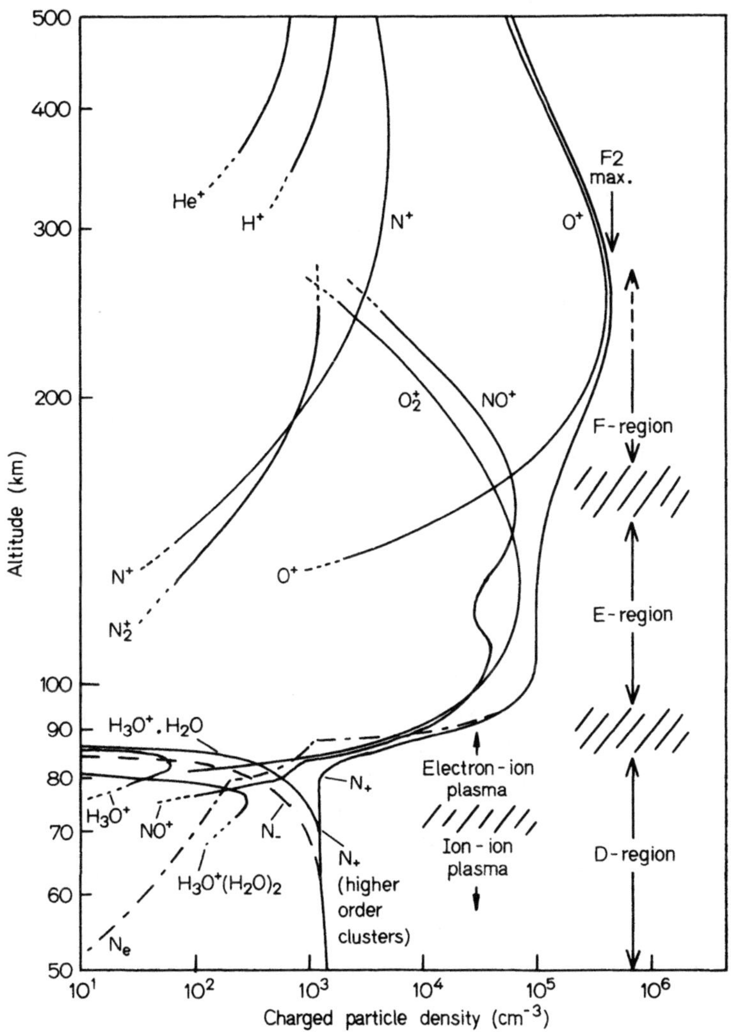

Fig. 4. The ionized regions of the Earth's atmosphere. The F, E and D regions are designated according to the "ledges" observed in the electron density. Typical altitudinal profiles of the various positive ion densities in the positive ion-electron plasma (from Refs.[20] and [21]) are also shown. The negative ion types in the positive ion-negative ion plasma of the lower D-region are known but the detail altitudinal profiles of density are not well characterised and so only the approximate total negative ion density, N_-, (dashed line) as obtained from Refs.[22] and [23]) is shown. The profiles of the electron density, N_e, and the total positive ion density, N_+, are also included. It is assumed that quasi-neutrality exists throughout the atmosphere, that is $N_e \approx N_+$ in the thermosphere, $N_e + N_- \approx N_+$ in the mesosphere, and $N_- \approx N_+$ in the stratosphere and troposphere

density below this altitude does decrease due to the rapid dissociative recombination of electrons with the water cluster ions which are efficiently formed in this region (Sect. 3.2.3), the process of electron attachment (Sect. 3.2.2) is also sufficiently rapid to convert some of the free electrons to negative ions which do not efficiently reflect radio waves and therefore remained undetected. Thus above 80 km an electron-positive ion plasma exists, whilst below this altitude a gradual transition towards a negative ion-positive ion plasma takes place which is complete below an altitude of about 60 km. These gross features of the Earth's ionosphere have been confirmed using several other techniques such as rocket-borne electrostatic probes [24] and radar backscatter [25].

Several distinct regions of ionization exist (Fig. 4) which have characteristic and predictable shapes, resulting from the interaction of a specific wavelength or a broad wavelength band of ionizing radiation with an atmosphere of varying composition. They have been designated the F, E and D layers (or regions) in order of descending altitude and as can be seen from Fig. 4, the electron density ranges from about $10^6 cm^{-3}$ at the F2 maximum to zero in the lower D-region. The ion and neutral compositions as well as the particular solar radiations which are most strongly absorbed in these regions have been determined using rocket-borne mass spectrometers and radiation detectors. Above the F2 maximum, the ionization results predominantly from the action of EUV and X-rays on the dominant neutrals (see Fig. 3) producing the observed ions He^+, N^+ and O^+ [20], that is , there is a correspondence between the ambient ions and neutrals. Ionic reactions however, are occurring, for example the nearly *thermoneutral charge transfer reaction* [26]

$$H^+ + O \longrightarrow O^+ + H \tag{3}$$

will be continuously influencing the relative concentrations of H^+ and O^+ ions. Below the F2 maximum the ion-to-neutral correspondence disappears and although N_2^+ and O^+ are expected to be the major primary ions, since *molecular nitrogen* is by far the most abundant neutral and the molecular oxygen is largely dissociated, the most abundance ion is NO^+ even though neutral NO is present in only an insignificant concentration (Fig. 1). As is now well known [20, 27], this is the result of the ionic reactions which occur such as the charge exchange and ion-atom interchange reactions:

$$\begin{aligned} N_2^+ + O &\longrightarrow NO^+ + N \\ &\longrightarrow O^+ + N_2 \end{aligned} \tag{4a}$$

$$O^+ + N_2 \longrightarrow NO^+ + N \tag{4b}$$

$$N_2^+ + O_2 \longrightarrow O_2^+ + N_2 \tag{4c}$$

So the chemistry of the F-region is generally well-understood and detailed chemical models have been constructed which can predict reasonably well the observed electron density and ionic composition [28, 29]. These models are based on laboratory data for ion-neutral reactions such as (4), and dissociative recombination reactions (Sect. 3.2.4), such as

$$NO^+ + e \longrightarrow N + O \quad (5a)$$

$$O_2{}^+ + e \longrightarrow O + O \quad (5b)$$

which control the loss of ionization in these regions. Many excellent reviews have been written on this subject [30–32]. However several intriguing questions still remain especially with respect to the rôle of metastable excited states of both neutrals and ions and of doubly-charged ions and this will be referred to in Sect. 3.2.2 and 4. The possible importance of metastable ions, e. g. O^+ (^2D), has been discussed from time to time [33–36] and they have recently been detected in the ionosphere (Sect. 3.2.2). The radiations in the E and F regions are sufficiently energetic to produce some doubly-charged species and it is worthy of note that the Atmosphere Explorer C (AE-C) satellite has recently detected O^{2+} in the F-region [37]. Such observations together with reliable laboratory data relating to excited ion [38–41] and doubly-charged ion [42–45] reactions (such data are just beginning to appear in the literature) will clarify the role of these species in ionospheric chemistry.

In the E-region where the most energetic solar UV and X-ray photons have been filtered out, the ions which are most efficiently produced are $O_2{}^+$ and O^+, the latter being quickly converted to NO^+ *via* reaction (4b). Hence the most abundant ions observed are $O_2{}^+$ and NO^+ (Fig. 4) which are relatively unreactive with the ambient neutrals present at these altitudes and so they persist until they are lost via reaction (5). However, this chemically less interesting region exhibits an unusual feature in the mid-latitude zones, the so-called "Sporadic-E" layers of enhanced ionization density within a layer a few kilometres thick is enhanced several-fold above the ambient ionization. Rocket-borne mass spectrometers have shown that *metallic ions* (Mg^+, Na^+, Si^+ and Fe^+) are concentrated in these layers[46] and more recently rocket flights have shown that metallic ions are also plentiful in the D-region [21,47, 48]. It seems that the metallic species originate from ablation of meteors in the atmosphere at D and E-region altitudes [48, 49]. Some data are available relating to metal ion reactions [50–54] although this is an area where more work is required.

In the D-region, the solar radiations capable of ionizing the major constituents of the region, ie. N_2 and O_2 have been filtered out and but for a fortuitous combination of circumstances, the ionization density in the D-region would be much smaller than actually exists ($\sim 10^3 - 10^4$ cm^{-3}). The existence of windows in the O_2 absorption spectrum allows H Lyman-α and H Lyman-β radiation, (intense lines in the solar radiation spectrum), to penetrate to the D-region and photoionize the minor neutral constituents NO and O_2 ($^1\Delta_g$) respectively (see Fig. 1). The metastable O_2 ($^1\Delta_g$) is produced by photo-excitation of O_2 ground state and by photodissociation of O_3 by long wavelength radiation which can also penetrate to these altitudes. Thus the primary positive ions will be predominantly NO^+ and O_2^+.

The technology of mass spectrometric sampling of the relatively high pressure D-region is more challenging than that for higher altitudes. However many of the problems have been solved and the in-situ experiments carried out by Narcisi and his colleagues [55, 56] have produced entirely unexpected results. Large concentrations of the *hydrated hydronium ions* $H_3O^+(H_2O)_n$, often called water cluster ions, were detected which in later, more refined experiments (designed to minimise the breakup of these weakly bound species during the sampling procedure) have been shown to dominate the positive-ion content of the lower D-region [47, 57]. More recent experiments by Arnold and his colleagues [58–61] have confirmed the dominance of these cluster ions (n is seen to be $\lesssim 20$) and any doubts which exist regarding the nature of the dominant D-region positive ions are ones of detail, i. e., what are the fractions of the various hydrates in a given altitude region and are the larger hydrates being satisfactorily sampled without collisional destruction, etc.

The origin and behaviour of these cluster ions has been the basis of much of the work on atmospheric ion chemistry during the last decade. The problem which is still not resolved in detail, is to explain how the primary ions NO^+ and O_2^+ are converted to the observed water clusters, i. e.

$$NO^+, O_2^+ \longrightarrow H_3O^+ (H_2O)_n \qquad (6)$$

always remembering that water vapour is a very minor constituent of the atmosphere at these altitudes (mixing ratio $\sim 10^{-6}$). Due to the collective effort of several laboratory groups [63–69] and aeronomists [70–72] (notably E. E. Ferguson and his colleagues at NOAA, Boulder, Colorado) a clear insight into the likely reactions involved has been obtained, and this will be discussed in Sect. 3 and 4. D-region chemistry has been the subject of several reviews [8, 73–75] and it is sufficient to say here that both the conditions of high pressure and low temperature in the D-region (and indeed in the stratosphere and troposphere) are conducive to ternary association reactions and to the production of weakly-bonded association ions viz:

$$O_2^+ + O_2 + N_2 \rightleftharpoons O_2^+ . O_2 + N_2 \qquad (7)$$

Such reactions, together with binary "switching" reactions of the kind

$$O_2^+ . O_2 + H_2O \longrightarrow O_2^+ . H_2O + O_2 \qquad (8)$$

have been shown in laboratory experiments [76] to be very effective in promoting the overall reaction indicated by reaction sequence (6).

A distinctive feature of the D-region is the occasional appearance of the highest cloud formation known to exist in the Earth's atmosphere. These so-called *"noctilucent clouds"* occur at or near the mesopause. Goldberg and Witt [48] using rocket-borne mass spectrometers have shown that the positive ions in these clouds are heavily clustered with water molecules and that hydrated hydronium ions are prominent. However, in addition, other hydrated species are present with masses that suggest that the core ions are Fe^+, FeO^+, FeO_2^+ and perhaps Mg^+ also, the metals presumably being of meteoritic origin. The co-existence of cluster ions with visible cloud formations adds some support to the view that the cluster ions play an important role as nucleation sites for droplet (aerosol) formation [77].

In the upper reaches of the D-region, electrons are the dominant negatively-charged species and electron-ion dissociative recombination (e. g. Eq. 5) controls the loss of ionization from the plasma, this process being considerably more rapid when cluster ions are involved (Sect. 3.2.4). However, at lower altitudes electrons are gradually replaced by negative ions until below about 60 km (70–80 km during nighttime), free electrons represent a negligible fraction of the ionized component of the atmosphere plasma. Loss of ionization from the ion-ion plasma is then due to the process of ion-ion mutual neutralization:

e. g., $H_3O^+(H_2O)_n + NO_3^-(H_2O)_m \longrightarrow$ products (9)

a process which is discussed in Sect. 3.2.5.

As is suggested by reaction (9), the ambient negative ions also are complex clusters. In-situ measurements of the negative ion composition have been made by Narcisi et al. [47] and Krankowsky et al. [58] and whilst there is no doubt that water clusters dominate, there remains some doubt as to the nature of the 'core' ions, i. e. whether the ions are predominantly $CO_3^-(H_2O)_m$ or $NO_3^-(H_2O)_m$. More of these very difficult sampling experiments are necessary to clarify the situation.

As in the case of the positive ion chemistry, the problem is to describe the chemical steps which convert the primary negative ions to the observed clusters. The primary negative ions can only be O^- and O_2^- formed in the electron attachment reactions:

$$e + O_2 + M \longrightarrow O_2^- + M \qquad (10a)$$

$$e + O_3 \longrightarrow O^- + O_2 \qquad (10b)$$

So the negative ion chemistry is summarised by

$$O_2^-, O^- \longrightarrow NO_3^-(H_2O)_m, CO_3^-(H_2O)_m \qquad (11)$$

which involves a number of parallel and sequential binary and ternary reactions. Some of the reactions considered to be involved will be discussed in Sect. 3.2.

Solar radiation sufficiently energetic to photoionize the ambient gas in the stratosphere cannot penetrate the atmosphere above it and the ionization in these regions is generated throughout the day and night by galactic cosmic rays. Thus the primary ions are O^+, O_2^+, N^+ and N_2^+, characteristic of the major atmospheric constituents O_2 and N_2. In-situ ion composition measurements in the stratosphere are fraught with difficulties and are only just beginning. The first observations were made by Arnold et al. [78, 79] again using rocket- and balloon-borne mass spectrometers. These revealed, as expected, a large number of different positive ions including the seemingly inevitable hydrated hydronium ions but with other clusters of the kind $X^+.(H_2O)_n$, the core ions, X^+, having not yet been convincingly identified. On the basis of the Arnold et al. data and the proton affinity data of Kebarle [80], Ferguson [81] has reasoned that some of the observed stratosphere-ions are of the type $(NaOH)_m H^+(H_2O)_n$. The recent balloon-borne mass spectrometer experiments of Olsen et al. [82] suggest that $H_3O^+(H_2O)_n$ ions dominate the positive ion content of the stratosphere down to 28 km. A similar conclusion has been arrived at by Arijs et al. [83] from their balloon flights but they also observed other ions some of which have masses corresponding to the proposed $(NaOH)_m H^+(H_2O)_n$ hydrates.

The negative ion content of the stratospheric plasma is even less certain. The only data so far available are those obtained very recently by Arnold [62] which are as yet preliminary but which show that the negative ions are generally more massive than the positive ions. Several negative ions within the mass range 81 ± 3 to 295 ± 3 amu were apparent in the spectra but the nature of these ions has not been ascertained. Arnold speculates that they may be of the kind $NO_3^- \cdot X_l \cdot Y_m Z_n$ with $l + m + n \leqslant 3$ and that X, Y and Z could be the acids HNO_3, HCl and H_2SO_4 which are known stratosphere neutral species, but it seems probable that HSO_4^- is an important core ion (see Sect. 3.2.3). It seems likely that the rich neutral chemistry which is occurring in the stratosphere and which generates reactive species such as the oxides of nitrogen etc.[84] will be intimately connected with and greatly influence the ion chemistry.

Little is known about the ion chemistry of the *troposphere.* Ionization is created by the action of galactic cosmic rays and in addition in the lower troposphere by radioactive emanation from rocks, both of which ensure that an appreciable degree of ionization exists at all times (Fig. 4). Thus the primary ions will be O^+, O_2^+, N^+ and N_2^+. No information is yet available concerning the actual ion composition of the troposphere because of the great experimental difficulties encountered in mass spectrometric sampling of such a high pressure gas supporting such a small fraction of ions, and so one can only speculate. However it seems certain that several species of both positive and negative ions will exist, all being clustered. The core positive ions might be predominantly H_3O^+ and $NaOH_2^+$ and the core negative ions NO_3^- or perhaps Cl^-. H_2O will probably be the most important cluster molecule (ligand), principally because of its high vapour pressure in the troposphere. However, other cluster molecules cannot be ruled out, for example, NH_3 in a sufficiently large concentration will displace H_2O from the larger $H_3O^+ (H_2O)_n$ ions (Sect. 3.2.3). NH_3 is continuously being introduced into the troposphere by biogenetic activity but it is readily dissolved by rain which is therefore an efficient removal process. Similarly HNO_3 strongly bonds to NO_3^-, but in the presence of large concentrations of water vapour, the equilibrium

$$NO_3^-(HNO_3)_n \overset{H_2O}{\rightleftharpoons} NO_3^-(H_2O)_m \qquad (12)$$

will be to the right. Thus local atmospheric conditions such as humidity, concentration of industrial pollutants and sunlight intensity, that is, all those factors which influence the neutral chemistry (and which in turn leads to undesirable environmental effects such as smog basins) will also influence the ion chemistry and hence to some extent determine the nature of the ambient ions.

The processes by which ions are lost in the stratosphere and the troposphere are not completely understood due to a sparcity of laboratory data on ionic recombination. It is most likely that mutual neutralization of cluster ions [reaction (9)] will be the primary loss mechanism in the upper stratosphere, with the process of collision-enhanced (ternary) recombination becoming increasingly important at lower altitudes (Sect. 3.2.5). In the presence of aerosols (liquid or solid droplets), loss of both positive and negative cluster ions from the gas phase can occur by attachment to the aerosol surfaces[85, 86] (see Sect. 4).

3 Elementary Charged-Particle Reactions of Significance in the Environmental Plasma

Ionization of a gas mixture with as many components as the Earth's atmosphere initiates a very complex ion chemistry. Chemical evolution, involving several reaction processes and very many individual reactions, eventually results in a dynamic equilibrium amongst the components of the medium which is both spatially and temporally variable. A detailed knowledge is required of the various reaction processes which can occur between electrons, positive and negative ions and neutral species in a low temperature plasma before the overall chemical evolution of the plasma can be adequately described. Therefore this section is largely concerned with recent advances in the laboratory studies of individual reaction processes and their relevance to the plasma chemistry of the various atmospheric regions. First, however, a general picture is presented of the overall chemistry of a low temperature gaseous plasma.

3.1 General Considerations

The general picture of chemical evolution in a low temperature plasma involving the charged particles is shown in Fig. 5. The ionizing radiation initially creates positive ions, electrons and neutral fragments such as atoms and free radicals in the gas. The positive ions will in general possess a wide distribution of internal energy states and may include multiply charged ions. The charged particle reactions which follow and their rates depend on the composition of the gas and on the existing physical conditions. The process of electron-ion recombination [e. g. reaction (5)] will occur immediately to some extent, which will tend to return the medium to the neutral (de-ionized) state. In the presence of electronegative gases, free electrons will also be lost via the process of electron attachment [e.g., reaction (10)] which generates negative ions. The additional charged-particle loss process of ionic reconbination [e. g., reaction (9)] is then possible. However, before the positive and negative ions are eventually lost by recombination reactions, a wide range of binary and ternary ion-neutral reactions will usually have occurred producing a variety of ions which are totally different in character from the primary ions. These reactions can include both fast charge transfer reactions of both ground state and excited state ions and sequential ternary association reactions [e. g., reaction (7)] which can build-up large cluster ions (both positive and negative). As is indicated in Fig. 5, at any point in the reaction chains, the ions can be lost from the plasma by recombination reactions and via these reactions the neutral composition as well as the ion composition of the medium is modified. Thus it is clear that a large amount of data relating to individual reactions are required, especially the rate coefficients and product ion distributions and their temperature dependences, if the most important reaction channels in the production of the ions observed in a plasma are to be identified. The more recent developments in this area of research are discussed in Sect. 3.2.

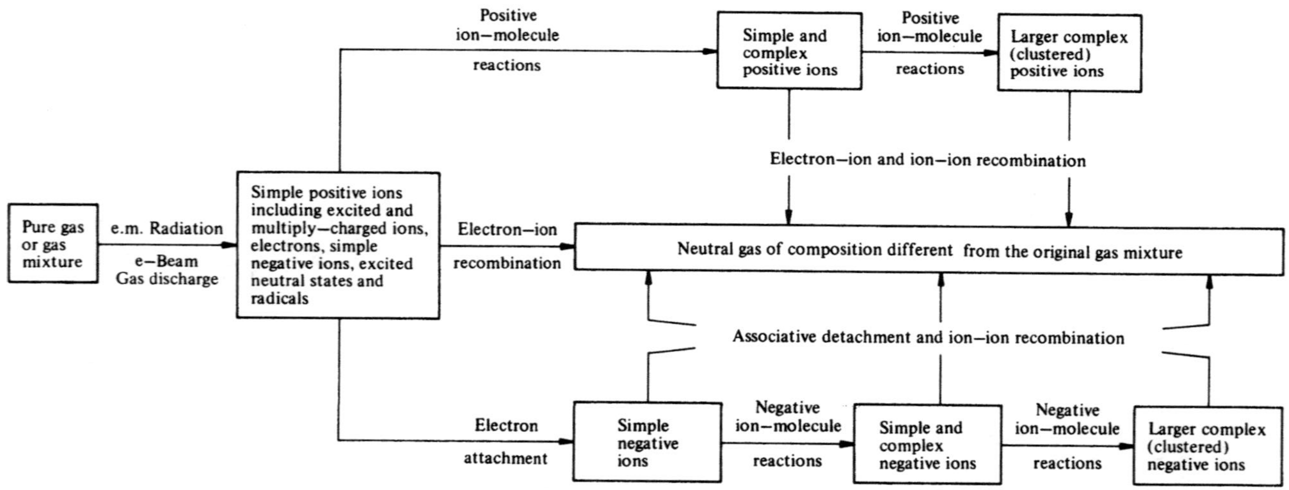

Fig. 5. Chemical evolution in a low temperature plasma. Illustrated are the types of charged particle reaction processes which may occur when ionization is created in a gas and which are terminated by recombination reactions, these returning the plasma to the neutral state

3.2 Individual Reaction Processes

It is pertinent to first briefly discuss the experimental techniques which have been applied to the study of the individual charged particle reaction processes before proceeding to highlight the most important laboratory measurements and their relevance to environmental plasma chemistry.

3.2.1 Experimental Techniques

Detailed descriptions of the various techniques are not appropriate here and only the salient features will be mentioned together with references to papers and reviews from which details can be obtained. The most appropriate data has been obtained using collision-dominated experiments (as opposed to ion beam and ion cyclotron resonance (ICR) techniques) in which the translational energies of the ions and electrons can be allowed to reach thermal equilibrium with the reactant charged or neutral particles before measurements are made. Also in such experiments, any internal excitation of the ions is effectively dissipated in collisions, except for specific metastable excited states which can survive and complicate data interpretation. In some cases the reactions of such excited states (which are now thought to have some ionospheric significance) have been successfully studied (Sect. 3.2.2). However, in general, the kinetic and internal energy states of the reactants in collision dominated experiments are clearly defined, i. e., a temperature can be ascribed to the system, and the data are directly applicable to the environmental plasma (in as far as its temperature can be ascertained).

Significant contributions to the field have been made using stationary afterglows (SA), flowing afterglows (FA), high pressure mass spectrometers (HPMS), drift tubes (DT), flow-drift tubes (FDT) and selected ion flow tubes (SIFT). Smaller, but significant, contributions have also been made using ion beams and the low pressure ion cyclotron resonance (ICR) techniques, but these will not be discussed here (for reviews on these techniques, see for example the book by McDaniel et al.[87)]). It is worthy of note that ICR has been successfully used recently to accumulate data relevant to the ion chemistry of interstellar molecular clouds[88)].

Some of the earliest measurements specifically directed towards gaining an understanding of the ionospheric plasma were made by Sayers and Smith and their colleagues[89–92)] using the SA technique. This technique involved the mass spectrometric sampling of the time-varying ionic wall currents in the afterglow of a pulsed r. f. discharge through an appropriate gas mixture. Later Lineberger and Puckett[64, 93, 94)] used a variant of the SA to study both positive and negative ion-molecule reactions. Smith and his co-workers[95–98)] also used the SA in conjunction with the Langmuir probe diagnostic technique to study a wide range of plasma processes including electron-ion recombination[99–101)]. However by far the most comprehensive study of electron-ion recombination (Sect. 3.2.4) has been made by Biondi and co-workers[102–104)] using the SA-microwave cavity diagnostic technique. These SA techniques have been reviewed by McDaniel and Mason[105)].

The HPMS technique developed and exploited by Kebarle and co-workers [106, 107] is very reminiscent of the SA technique. Ionization is created in a pure gas or gas mixture either by using a radioactive source [108] or a high energy electron beam [109]. Ions are sampled by a pin-hole orifice in the walls of the chamber and reaction rate coefficients deduced. A great deal of data relating to ternary association reactions (Sect. 3.2.3) has been obtained using this technique.

The development of the FA technique by Ferguson, Fehsenfeld and Schmeltekopf [110–113] resulted in an explosive increase in the amount of data relating to ionospheric chemistry. Ionization is generated in a flowing gas in a cylindrical tube about 10 cm diameter and about 1 metre long and the ions (positive or negative) in the afterglow plasma (which is distributed along the length of the flow tube) are sampled using a pinhole orifice/mass spectrometer combination located downstream. Reactant gases are introduced into the thermally-relaxed afterglow plasma downstream and hence remote from the source of ionization which thus avoids internal excitation of the reactant gas molecules (unavoidable in the SA). The versatility of the FA has enabled a wide variety of both positive and negative ions to be created in the flowing plasma and thus a very large number of binary and ternary reactions of environmental importance to be studied, including several over an appreciable temperature range [113–115]. Thus it has resulted in a major contribution to the understanding of ion-neutral reactions at thermal energies and to the elucidation of the ion chemistry of the ionosphere. Recently, Smith and his co-workers [116, 117] have combined their Langmuir probe diagnostic technique with the FA to study several plasma reaction processes of environmental interest including binary ion-ion mutual neutralization reactions of both simple and clustered ions [118–120] (see Sect. 3.2.5).

The temperature range over which critical data is required for the purpose of environmental chemistry is not large (Sect. 2.1) and the temperature variable SA, FA and HPMS have provided much valuable data. However, the maximum temperature accessible to these techniques is limited and so variants of the DT technique have been used to study rate coefficients as a function of ion kinetic energy (up to several electron volts). The temperature dependences of the rate coefficient are then deduced (not without difficulty) from the data. The DT technique of Biondi and his co-workers [121–123] and the FDT technique, developed and carefully exploited by Albritton and others [124, 125], which combines the versatility of the FA with the conventional DT technique, have proven to be especially useful (Sect. 3.2.2). The HPMS technique developed by Castleman and his co-workers [126, 127] is also providing valuable data of environmental interest.

The very recent development by Adams and Smith [128, 129] of the SIFT technique is providing a large amount of data on both binary and ternary ion-molecule reactions over the approximate temperature range 100–600 K [130–133]. In essence, the technique involves the injection at low energy of a mass-selected positive or negative-ion beam into a flowing gas. Virtually any positive or negative ion, including metastable excited ions, doubly charged ions and cluster ions, originating from any type of ion source, can be introduced into a flow tube and rate coefficients and product ion distributions determined for their reactions with any gas or vapour including unstable species (e. g. atoms and radicals). Details of the technique have been given in a recent review [134]. The versatility of the SIFT is such that it is being exploited to study a wide range of interesting problems and has already had notable success in providing critical data relating to molecular synthesis in interstellar clouds [135, 136]. The significant amount of data obtained to date with the SIFT relating to atmospheric chemistry (Sect. 3.2.2 and 3.2.3) will undoubtedly be considerably increased in the next few years. This technique and the others discussed in this section are continuing to be developed and exploited in attempts to solve the increasingly complicated problems being met in the ion chemistry of the stratosphere and the troposphere.

3.2.2 Binary Ion-Molecule Reactions

Positive ion binary reactions are exemplified by reaction (4). They are important in all regions of the atmosphere and dominate the ion chemistry of the E- and F-regions of the ionosphere. The ion chemistry of the E- and F-regions has been clear for sometime[30, 31], essentially involving a few binary positive ion-molecule reactions which lead finally to the two terminating ions NO^+ and O_2^+, which are eventually lost by dissociative recombination (reaction 5). Thus the recent work has been concerned largely with obtaining more precise laboratory data on the few most important reactions, that is in determining their branching ratios (product ion distributions) and the temperature dependences of their rate coefficients, so that more detailed modelling of the chemistry of the ionosphere can be carried out. The rôles of excited ions and doubly charged ions are also being considered as reliable laboratory data on their reactions and in-situ data on their concentrations in the ionosphere become available (see below).

Amongst the first ionospherically important reactions to be studied were:

$$He^+ + N_2 \longrightarrow N^+ + N + He \ (60\%) \tag{13a}$$

$$\longrightarrow N_2^+ + He \quad (40\%) \tag{13b}$$

$$He^+ + O_2 \longrightarrow O^+ + O + He \ (\gtrsim 98\%) \tag{14a}$$

$$\longrightarrow O_2^+ + He \quad (\lesssim 2\%) \tag{14b}$$

Their rate coefficients, k, were first measured at 300 K in the SA [91] and were both found to be gas kinetic ($k \sim 10^{-9} cm^3 s^{-1}$). Subsequent FA studies at 300 K [137] confirmed these results. More recent FA measurements [112, 113] have shown that k is independent of temperature over the range 80–900 K which, on the available evidence, seems to be a general feature of fast ion-molecule reactions. The energy independence of such reactions is also evident from DT and FDT studies [121, 138]. The earlier SA and FA measurements did not provide the product distributions for reactions (13) or (14), but these have since been determined at 300 K using several techniques[121, 129, 139–143], and very good agreement has been obtained for reaction (13) for which the dissociative charge transfer channel leading to N^+ is favoured (60% as shown). However reaction (14) has given more problems. The recent SIFT[129] and DT[143] data both indicate that the dissociative channel (14a) is almost totally dominant ($\gtrsim$ 98%) whereas the earlier FA[39] and DT[121] measurements gave an erroneously high proportion of O_2^+. This is because the determination of the product distribution for reaction (14) is complicated by the occurrence of the secondary reaction

$$O^+ + O_2 \longrightarrow O_2^+ + O \quad (15)$$

which, if not accounted for, leads to an overestimate for the O_2^+ channel. Secondary reactions are always a potential problem in the determination of product distributions and are more serious the faster the secondary reactions. The interesting point in this case is that reaction (15) for ground state O^+ ions is known to be relatively slow ($k = 2 \times 10^{-11}$ $cm^3 s^{-1}$) and this misled the earlier workers[121, 139] who therefore assumed that the production of O_2^+ via reaction (15) would be slow. Johnsen et al.[143] have reasoned that the earlier DT data indicates that the O^+ produced in reaction (14) is in an excited state which reacts more rapidly with O_2 than does ground state O^+. Recent (unpublished) SIFT data has confirmed this postulate; it appears that a substantial fraction of the O^+ is formed in a metastable state. The SIFT data of Glosik et al.[39] has shown that the rate coefficient for the reaction of metastable $O^+(^2D, ^2P)$ ions with O_2 is considerably larger than for ground state O^+ ions.

Reaction (15) and the reaction

$$O^+ + N_2 \longrightarrow NO^+ + N \quad (16)$$

are both important in E and F region ion chemistry in that they convert atomic ions to molecular ions and therefore enhance the electron-ion recombination loss rate (Sect. 3.2.4). At 300 K reactions (15) and (16) are both relatively slow (k for reaction (16) is 1.2×10^{-12} $cm^3 s^{-1}$) and experience has shown that the rates of reactions with small rate coefficients at 300 K are likely to vary with temperature. Smith and Fouracre[144] using a SA first showed that for reaction (15), $k \propto T^{-1/2}$ and this was subsequently confirmed by both FA[112, 113, 145] and temperature variable DT[123] studies. Reaction (16) was studied in detail as a function of temperature[112, 113, 123, 145] and ion kinetic energy[125] after Schmeltekopf et al.[146] had shown that its rate coefficient increased dramatically with increasing vibrational temperature of the reactant N_2. A correspondingly rapid increase in the rate coefficient is observed with increasing O^+ kinetic energy[125].

The temperature and energy dependence of the rate coefficients for most of the ground state ion-molecule reactions of E- and F-region significance have now been adequately determined[147], including that for the reaction

$$N^+ + O_2 \longrightarrow NO^+ + O \quad (50\%) \quad (17a)$$

$$\longrightarrow O_2^+ + N \quad (44\%) \quad (17b)$$

$$\longrightarrow O^+ + NO \quad (6\%) \quad (17c)$$

It is interesting to note that the earlier FA experiments only recognised the two major product channels in reaction (17) whereas the SIFT, which is especially valuable for the determination of product distributions, identified the minor O^+ product channel (which has since been observed in an ICR experiment[88]). The acquisition of such detailed data, especially temperature dependence data, is essential for the detailed modelling of the E- and F-region ion chemistry, since in these regions the temperature changes markedly with altitude (Fig. 2).

In most chemical models of the ionosphere, little attention has been paid to the possible rôle of excited ions, although as early as 1963, Dalgarno and McElroy[33, 148] had discussed the likely presence of significant concentrations of excited ions in the ionosphere and calculated their production rates by solar photons (see also the more recent review by Dalgarno[149]). The presence of excited ions, specifically $O^+(^2P)$ and $O^+(^2D)$, has now been confirmed in the F-region by instruments carried aboard the Atmosphere Explorer C and D Satellites (AE-C and AE-D)[150] and these observations, coupled with the available data on the ionic and neutral composition of the thermosphere, have stimulated renewed activity in this field. Rate coefficients and quenching (de-excitation) coefficients have been estimated from the satellite data for the reactions of $O^+(^2D)$ and $O^+(^2P)$ with O, O_2 and N_2[34–36]. At the time that these estimates were obtained, no laboratory data were available for comparison but recently a good deal of data has been obtained on excited ion reactions of ionospheric interest using the SIFT technique[39, 40] which includes measurements of the rate coefficients, k, for several $O^+(^2P, ^2D)$ reactions, e.g.

$$O^+(^2P, ^2D) + O_2 \longrightarrow O_2^+ + O \tag{18}$$

k has also been measured for reaction (18) by Mauclaire et al.[151], who obtained a value of 6.5×10^{-10} cm^3 s^{-1}, in conflict with that given in Ref. [39]. However, it has recently been brought to our attention by D. L. Albritton that the values of k for reaction (18) and for the corresponding N_2 reaction reported in Ref.[39] are erroneously low (this we have now shown to be due to computational errors). Thus using our SIFT we have re-measured k for reaction (18) to be $(6.5 \pm 2) \times 10^{-10}$ cm^3 s^{-1} at 300 K and that for the $O^+(^2P, ^2D) + N_2$ reaction to be $(7.5 \pm 2) \times 10^{-10}$ cm^3 s^{-1} [215]. These are in good agreement with the values obtained by Albritton et al.[216] who measured k for both reactions to be $(8.35 \pm 3.4) \times 10^{-10}$ cm^3 s^{-1}, and with the value for the O_2 reaction obtained by Mauclaire et al. Similar results have also been obtained recently by Johnsen and Biondi[217] for these reactions. In none of these laboratory experiments has the state (or states) of the O^+ ions involved yet been identified. The laboratory data can now be compared with the estimates from satellite observations[36]. For the reactions

$$O^+(^2D) + N_2 \longrightarrow N_2^+ + O \tag{19a}$$

$$O^+(^2D) + O_2 \longrightarrow O_2^+ + O \tag{19b}$$

k is estimated to be $(5 \pm 2.5) \times 10^{-10}$ cm^3 s^{-1} for (19a) and $(2 \pm 1.5) \times 10^{-9}$ cm^3 s^{-1} for reaction (19b).

Detailed chemical models of the thermosphere have been produced by Oppenheimer et al.[29] based on the satellite observations, which include estimated rate coefficients for several excited ion reactions. Much of the ionic reaction rate data derived from the AE satellite experiments has recently been reviewed by Torr and Torr[36].

In addition to metastable O^+ ions, other excited ions will surely be generated in the ionosphere. The ions O_2^+, NO^+, N_2^+ and N^+ all have long-lived metastable electronic states, but to date they have not been observed in the ionosphere. A considerable amount of laboratory data is now becoming available on the reaction rate coefficients of these excited ions. Of note are the FDT[38, 41] and SIFT[39] data on O_2^+ (a $^4\Pi_a$) and NO^+ (a $^3\Sigma^+$) reactions and the SIFT data on the reactions of metastable electronic states of N^+ and N_2^+ [40]. These excited ions usually react rapidly with most molecules, direct charge transfer is the most favoured channel but collisional quenching of the excited ion [e.g. reaction (19b)] is often a significant channel, especially for the NO^+ and O_2^+ excited ion reactions.

Doubly charged ions, specifically O^{2+}, have recently been detected in the thermosphere by the AE-C satellite. The photochemistry of this species has been discussed by Brieg et al.[37] who has also estimated the rate coefficients for the reactions

$$O^{2+} + O \rightarrow O^+ + O^+ \qquad (20)$$

$$O^{2+} + N_2 \rightarrow \text{products} \qquad (21)$$

to be $\sim 1 \times 10^{-11}$ cm^3 s^{-1} and $< 1 \times 10^{-11}$ cm^3 s^{-1} for reactions (20) and (21) respectively. These estimates seem rather low on the basis of the little previous data available for doubly charged ions[42] and further doubt has been cast on their validity by the recent DT measurements of Johnsen and Biondi[45], which indicate that the rate coefficients for the reactions of O^{2+} with both N_2 and O_2 are fast ($k \sim 1 \times 10^{-9}$ cm^3 s^{-1}). A possible explanation of the low values derived from the satellite work is given in a recent paper[218]. It is worthy of note, however, that the recent SIFT data for the reactions of the ground and excited metastable states of Xe^{2+} and Ar^{2+} with several atomic and molecular gases indicates a great variety of mechanisms and rates for such reactions[43, 44]. Single and double charge transfer and collisional quenching of the excited states variously (and unpredictably) occur and the rate coefficients, while usually large, are in some cases very small. A good deal of accurate laboratory data is needed before the rôle of doubly-charged ions in ionospheric chemistry can be ascertained. The likely rôle of excited-ion and doubly-charged ion reactions in ionospheric chemistry will be referred to briefly in Sect. 4.

As discussed in Sect. 2.2, the ionization in the D-region (mesosphere) is largely generated by the near-resonant photoionization of NO and O_2 producing NO^+ and O_2^+ in their ground electronic states. Therefore significant concentrations of excited ions are not generated in this region and the positive ion chemistry largely begins with the ternary association reactions of ground state NO^+ and O_2^+ ions (Sect. 3.2.3). At lower altitudes in the stratosphere and the troposphere where the primary source of ionization is cosmic rays, the primary ions will be the ground and excited states of N^+, O^+, O_2^+ and N_2^+. These species will then undergo fast binary and ternary reactions with the major and minor neutral constituents of the local atmosphere. The overwhelming abundance of N_2 and O_2 will ensure that the products of the binary reactions will again be the low recombination energy ions NO^+ and O_2^+ [via reactions (4b), (4c), (15) to (17)]. At the high pressures and low temperatures in the lower atmosphere, the ternary association reaction (7) will generate O_4^+ ions and the corresponding reactions

$$N^+ + N_2 + M \longrightarrow N_3^+ + M \qquad (22)$$

$$N_2^+ + N_2 + M \longrightarrow N_4^+ + M \qquad (23)$$

will generate N_3^+ and N_4^+. Under favourable conditions sequential ternary association reactions can generate larger associated ions (see Sect. 3.2.3). However the relatively energetic N_3^+ and N_4^+ ions can also undergo binary reactions, e.g.

$$N_3^+ + O_2 \longrightarrow NO^+ + O + N_2 \quad (70\%) \qquad (24a)$$

$$\longrightarrow NO_2^+ + N_2 \quad (30\%) \qquad (24b)$$

$$N_4^+ + O_2 \longrightarrow O_2^+ + N_2 + N_2 \qquad (25)$$

Note the generation again of NO^+ and O_2^+ as well as the ion NO_2^+. A detailed study has recently been reported of the reactions of the primary and secondary stratospheric ions with several molecules[152)] and of the reactions of O_4^+ and O_5^+ with several stratospheric neutrals[153)]. It seems clear from these studies that although fast binary ion-molecule reactions are important first steps in the positive ion chemistry of the lower atmosphere, the subsequent chemistry is controlled by ternary association reactions (Sect. 3.2.3).

Negative ions become an increasingly important component of the environmental plasma below an altitude of about 80 km where the pressures are sufficiently high to promote 3-body electron attachment, exemplified by the reaction

$$e + O_2 + N_2 \longrightarrow O_2^- + N_2 \qquad (26)$$

At lower altitudes where significant concentrations of ozone exist, O^- ions are generated by dissociative attachment [reaction (10b)]. These electron attachment processes and the laboratory techniques used to determine their rate coefficients were reviewed some time ago by Phelps[154)]. In the stratosphere and troposphere, negative ions can also be generated by dissociative attachment reactions of thermalised electrons with pollutants[155, 156)] such as the freons: e.g.

$$e + CCl_2F_2 \longrightarrow Cl^- + CClF_2 \qquad (27a)$$

$$\longrightarrow F^- + CCl_2F \qquad (27b)$$

Although reactions of the kind (26) always dominate negative ion production.

Recent laboratory studies have concentrated on the loss mechanisms for the primary negative ions. The process of associative detachment:

$$O^- + O_2(^1\Delta_g) \longrightarrow O_3 + e \qquad (28)$$

$$O_2^- + O \longrightarrow O_3 + e \quad (50\%) \qquad (29a)$$

$$(\longrightarrow O^- + O_2) \quad (50\%) \qquad (29b)$$

which has been studied almost exclusively by Fehsenfeld and his colleagues[157], is a very efficient loss process for negative ions in the D-region. For ground state O_2, reaction (28) is endoergic [the reverse reaction is then very efficient, see reaction (10b)]. Channel (29b) is not, of course, an associative detachment reaction but rather a charge transfer reaction which can proceed because the electron affinity of O atoms is greater than that of O_2 molecules. A knowledge of the electron affinities of atmospheric neutral species is necessary if the negative ion chemistry is to be elucidated Thus the determination of electron affinities has received considerable attention in recent years and the exploitation of the laser photodetachment technique by Lineberger and co-workers[158] is adding greatly to the body of critical data.

A reaction of particular interest in aeronomy has been

$$O_3^- + CO_2 \longrightarrow CO_3^- + O_2 \qquad (30)$$

since it could be an important step in the generation of CO_3^- $(H_2O)_n$ cluster ions which have been tentatively identified in the D-region. It has been shown[159] that when the O_3^- is hydrated [i.e. O_3^- $(H_2O)_n$] its reactivity with CO_2 rapidly decreases and the reaction may even become endoergic for large hydrates[160]. Thus in the lower atmosphere where the mixing ratio of water vapour is appreciable, hydration of O_3^- could prevent its reaction with CO_2. This illustrates how the reactivity of an ion can be greatly influenced by the addition to it of weakly bonded ligands, a phenomenon recognised also for positive ions[161, 162], and much more laboratory data is required in this area if the ion chemistry of the lower atmosphere is to be properly understood. Detailed schemes of D-region[8, 74, 75, 147] and stratospheric and tropospheric[147] ion chemistry have been proposed.

Whilst the rate coefficients for many binary and ternary negative ion-molecule reactions have been acquired recently, predominantly using the FA technique (see the data compilation of Albritton[115]), many more are required if the important paths in the synthesis of the observed negative ion clusters are to be identified. Product distributions have been studied even less for negative ion-molecule reactions, principally because of experimental difficulties, yet more than one product channel is accessible in several atmospherically important reactions[73], for example,

$$O_3^- + NO_2 \longrightarrow NO_2^- + O_3 \qquad (31a)$$

$$\longrightarrow NO_3^- + O_2 \qquad (31b)$$

The SIFT technique is especially suited to the determination of product distributions and these will undoubtedly be available for some of the relevant binary negative ion reactions in the near future. Hopefully, such data will help to solve several basic problems. For example, although it seems certain that NO_3^- is formed efficiently in the lower atmosphere (see Sect. 2.2), it is by no means certain how it is generated. Reaction (31b) may be an important route but it is unlikely to be exclusive. HNO_3 and N_2O_5, both stratospheric constituents react with most negative ions to produce NO_3^- [156, 163] but the most important paths have not yet been confidently identified. The rôle of minor neutral constituents is one of the major outstanding problems in the negative ion chemistry of the lower atmosphere (Sect. 4).

3.2.3 Ternary Ion-Molecule Reactions

As previously mentioned, ternary (or 3-body) ion-molecule reactions are only significant in the Earth's atmosphere below the mesopause (~ 80 km) where they play a crucial part in the ion chemistry. In this region of the atmosphere, the major problem to be solved is to determine the ionic reaction paths which convert the primary positive and negative ions to the dominant positively and negatively charged water cluster ions [see reactions (6) and (11)]. Most progress has been made in elucidating the positive ion chemistry, so this will be considered first.

In the D-region, the primary *positive ions* are NO^+ and O_2^+ (NO^+ being dominant) and so the problem reduces to one of finding a quantitative chemical scheme for the conversion of NO^+ to $H_3O^+(H_2O)_n$ ions, the steps of which are sufficiently rapid to remove NO^+ at a rate commensurate with its known production rate from photoionization of NO. The first proposal was made by Fehsenfeld and Ferguson[63] who, with their colleagues, have contributed so much to this area of research. Their scheme, based on their FA data, is summarised by the following reactions, the first step being the direct 3-body hydration of NO^+:

$$NO^+ + H_2O + M \longrightarrow NO^+ \cdot H_2O + M \qquad (32)$$

followed by

$$NO^+(H_2O)_{1,2} + H_2O + M \longrightarrow NO^+(H_2O)_{2,3} + M \qquad (33)$$

then

$$NO^+(H_2O)_3 + H_2O \longrightarrow H_3O^+(H_2O)_2 + HNO_2 \qquad (34)$$

The ternary reactions (32) and (33) sequentially build up the NO^+ hydrates until $NO^+(H_2O)_3$ is produced, which then undergoes a binary reaction to give the hydrated hydronium ions. Further increases in the size of the hydrates is considered to be via direct clustering:

$$H_3O^+(H_2O)_n + H_2O + M \longrightarrow H_3O^+(H_2O)_{n+1} + M \qquad (35)$$

In the atmosphere, M is an abundant neutral i.e. either N_2 or O_2. Whilst the authenticity of this chemical scheme is not in doubt, since it is based on a considerable amount of data from several groups[64, 66–69], model calculations have shown that it is not able to quantitatively explain the rate of loss of NO^+ in the ionosphere, the major problem being that reaction (32) leading to the first NO^+ hydrate, is not sufficiently rapid. The details of this problem have been discussed in several reviews[8, 164–166].

Alternative, more rapid paths for the production of $NO^+ \cdot H_2O$ have been proposed[167–170] involving as a first step the ternary association of NO^+ with the abundant atmospheric CO_2 and N_2

$$NO^+ + CO_2 + M \rightleftharpoons NO^+ \cdot CO_2 + M \quad (36)$$

followed by the fast binary switching reaction

$$NO^+ \cdot CO_2 + H_2O \longrightarrow NO^+ \cdot H_2O + CO_2 \quad (37)$$

Similarly

$$NO^+ + N_2 + M \rightleftharpoons NO^+ \cdot N_2 + M \quad (38)$$

$$NO^+ \cdot N_2 + H_2O \longrightarrow NO^+ \cdot H_2O + N_2 \quad (39)$$

Although the rate coefficients for reactions (36) and (38) were expected to be relatively small (based on the limited amount of laboratory data available[167, 171)]), it was suggested that the abundance of CO_2 and N_2 in the D-region would ensure that the loss rate of NO^+ via these reactions was adequately fast, and switching reactions of the kind (37) and (39) were known to be fast[66, 76)]. These schemes have been given greater credence by the in-situ observations of Arnold and Krankowsky[59)] of ions designated as $NO^+ \cdot N_2$ and $NO^+ \cdot CO_2$, as well as mixed clusters of the kind $NO^+ \cdot H_2O \cdot CO_2$ etc. Unfortunately at the onset meaningful quantitative assessment of the importance of these suggested schemes was hampered because the available measurements of the NO^+ ternary association reactions were quite inconsistent for a given reaction. This is principally because of the experimental difficulties inherent in the study of reactions of such weakly-bonded ions such as $NO^+ \cdot N_2$ which are efficiently collisionally dissociated in a reaction chamber at the temperatures at which the experiments are normally carried out. Thus the reverse reactions in (36) and (38) are appreciable and in order to determine the forward rate coefficients a "scavenger" gas such as H_2O has to be introduced into the reaction vessel, at a critical concentration which is difficult to determine, to convert the weakly bonded ion to the more stable $NO^+ \cdot H_2O$ (i.e. making use of reactions (37) and (39)). In a very careful experiment, Johnsen et al.[172)] have succeeded in obtaining the forward rate coefficient and the equilibrium constant for the reaction

$$NO^+ + N_2 + N_2 \rightleftharpoons NO^+ \cdot N_2 + N_2 \quad (40)$$

at three temperatures appropriate to the D-region (130, 180 and 220 K), and also deduced the bond energy in $NO^+ \ldots N_2$ (= 0.18 eV). The forward rate coefficient was seen to vary as $T^{-4.4}$. A similar temperature dependence ($\sim T^{-5}$) has been subsequently inferred from in-situ ionospheric observations[173)]. In a recent laboratory study, Smith et al.[130)] have determined the forward rate coefficients for reactions (36) and (38) at 225 K and 300 K, the most significant result from this study being the rapid temperature dependence ($\gtrsim T^{-6}$) for reaction (36). Using their data and that of Johnsen et al., Smith et al. have constructed a limited quantitative chemical scheme to describe the conversion of NO^+ to $NO^+ \cdot H_2O$ in the D-region which involves reactions (36) to (40), and concluded that both reactions (36) and (38) are important in generating the $NO^+ \cdot H_2O$, with reaction (36) becoming relatively more important at the lower D-region temperatures. More detailed D-region schemes have been constructed by Thomas[71)] and Reid[72)] but they do not include this most recent data.

Whilst the formation of the first NO^+ hydrate is considered to be an essential first step in the production of $H_3O^+(H_2O)_n$ ions, the subsequent steps are not so clear. Reactions such as (33) and (34) will probably contribute, but other reactions are expected to be involved. The in-situ observation of mixed clusters of NO^+ variously with H_2O, CO_2, N_2 and O_2 indicates that association reactions of the kind:

$$NO^+ \cdot H_2O + CO_2 + N_2 \rightleftharpoons NO^+ \cdot H_2O \cdot CO_2 + N_2 \quad (41)$$

are involved in the overall reaction scheme[59)]. No laboratory data is as yet available for such reactions, but with new advances in experimental technique, such as the temperature variable SIFT[134)], data on such reactions should soon be forthcoming.

In the higher pressure and low temperature environments of the stratosphere and troposphere, ternary association reactions can proceed very rapidly. Reactions (4b), (4c), (15–17), (22–25) rapidly convert the primary ions to O_2^+ predominantly but with smaller concentrations of NO^+ and NO_2^+ [via reaction (24b)]. These ions will then be converted to cluster ions via reaction schemes[65, 174)] analogous to the D-region schemes outlined above. The O_2^+ is first converted to the $O_2^+ \cdot H_2O$ via reactions (7) and (8), then the binary reaction sequence:

$$O_2^+ \cdot H_2O + H_2O \longrightarrow H_3O^+ + OH + O_2 \quad (42a)$$

$$\longrightarrow H_3O^+ \cdot OH + O_2 \quad (42b)$$

followed by:

$$H_3O^+ \cdot OH + H_2O \longrightarrow H_3O^+(H_2O) + OH \quad (43)$$

which has been observed to occur in FA experiments[175)], produces the hydrated hydronium ion. Larger clusters can then be generated as discussed above [reaction (35)]. With the greater number of minor neutral constitutents which exist in the atmosphere at lower altitudes (Fig. 1), the detailed ion chemistry will be very complex, probably involving mixed clusters of the common atmospheric constituents and minor constituents such as H_2CO, CH_3OH, NH_3 etc. as intermediates in the production of the terminating cluster ions. In the presence of significant concentrations of NH_3 or other species of low ionization energy, rapid binary transfer reactions can also occur such as[152, 176)]

$$O_2^+ + NH_3 \longrightarrow NH_3^+ + O_2 \quad (44)$$

$$O_2^+ + H_2CO \longrightarrow H_2CO^+ + O_2 \quad (90\%) \quad (45a)$$

$$\longrightarrow HCO^+ + HO_2 \quad (10\%) \quad (45b)$$

$$O_2^+ + CH_3OH \longrightarrow CH_3OH^+ + O_2 \quad (50\%) \quad (46a)$$

$$\longrightarrow H_3CO^+ + HO_2 \quad (50\%) \quad (46b)$$

The low recombination energy product ions can then undergo binary and ternary reactions which contribute to the establishment of a complex mixture of cluster ions. For example, subsequent to the production of NH_3^+ by reaction (44), the following reactings might occur:

$$NH_3^+ + H_2O \longrightarrow NH_4^+ + OH \tag{47}$$

and/or

$$NH_3^+ + NH_3 \longrightarrow NH_4^+ + NH_2 \tag{48}$$

followed by

$$NH_4^+ + H_2O + M \longrightarrow NH_4^+ \cdot H_2O + M \tag{49}$$

and/or

$$NH_4^+ + NH_3 + M \longrightarrow NH_4^+ \cdot NH_3 + M \tag{50}$$

Reactions (47) to (50) have all been studied in the laboratory[88, 159, 177]. It is also known from laboratory studies[159, 162, 177] that NH_3 can displace H_2O from small water clusters, e.g.

$$H_3O^+(H_2O)_{1,2} + NH_3 \longrightarrow NH_4^+(H_2O)_{0,1} + 2\,H_2O \tag{51}$$

but for relatively large H_2O to NH_3 concentrations (which is the likely situation in the atmosphere) the equilibrium in the reaction scheme

$$H_3O^+(H_2O)_W \underset{H_2O}{\overset{NH_3}{\rightleftharpoons}} \begin{bmatrix} H_3O^+(H_2O)_X(NH_3)_Y \\ NH_4^+(H_2O)_X(NH_3)_Y \end{bmatrix} \underset{H_2O}{\overset{NH_3}{\rightleftharpoons}} NH_4^+(NH_3)_Z \tag{52}$$

will be over to the left. Similar reaction schemes between water cluster ions and the high proton affinity species CH_3OH and H_2CO have been studied in the FA and their relevance to stratospheric ion chemistry considered[161]. Reactions of the kind:

$$H_3O^+(H_2O)_n + H_2CO \longrightarrow H_3CO^+(H_2O)_n + H_2O \tag{53}$$

occur producing protonated hydrates, which in the particular case of reaction (53) are hydrated protonated formaldehyde ions only when $n = 0, 1, 2$. For $n > 2$ reaction (53) does not proceed and the water clusters are stable against reaction with H_2CO. Clearly, the relative proton affinities of H_2O and H_2CO are important in determining the energetics of such reactions. As was mentioned in Sect. 2.2, Ferguson[81] has suggested that the ions observed by Arnold et al.[78, 79] in the stratosphere (Sect. 2.2) are hydrated protonated sodium hydroxide. This implies that reactions of the kind

$$H_3O^+(H_2O)_n + NaOH \longrightarrow (NaOH)H^+(H_2O)_n + H_2O \quad (54)$$

are occurring, which are energetically allowed because of the large proton affinity of NaOH[107]. Further detailed laboratory studies of reactions of this kind are necessary as well as more in-situ observations of stratospheric ions and neutrals before these ideas can be substantiated.

Although a great deal of data is not available concerning many of the association reactions of probable atmospheric significance, especially those involving the larger clusters, much can and has been gained by studying other clustering reactions which are more amenable to study with existing techniques. A body of data exists, largely from the laboratories of Kebarle[107] and of Castleman[126, 127, 178] on the rates of the reactions of a variety of cluster ions possessing various core ions (e.g. H_3O^+, Li^+, Sr^+ etc.) including critical thermodynamic data for such reactions and data on the bond strengths of the ligands as a function of the number of ligands clustered to the core ion. An especially interesting result from these studies is the apparent existence of "hydration shells", recognised by a significant change in the stability of the clusters above a given number of ligands within it[179]. Based on the available data and a detailed consideration of the thermodynamics of ion clustering, Castleman[180] has suggested that both positively and negatively charged cluster ions will probably act as efficient nucleation sites leading to aerosol formation in the atmosphere.

Until recently, little reliable data was available on the temperature dependence of ternary association reactions. Good[181] has reviewed the data available up to 1975. With the inception of the SIFT technique accurate temperature dependencies have been obtained for several ternary association reactions which indicates that the variation of the ternary rate coefficients with temperature closely conforms to a simple power law behaviour ($k \propto T^{-n}$) as predicted by statistical theory, but with n much smaller than predicted[131–133]. Such data is contributing to a growing understanding of the mechanistic aspects of ion-molecule association reactions[134, 135].

Negative ions are produced initially in the lower atmosphere via reactions (10b) and (26), but because of the high O_2 partial pressure reaction (26) is dominant and so most schemes describing the evolution of the negative ions in the atmosphere consider O_2^- as the primary ion. Important loss processes for O_2^- are considered to be the ternary association reactions

$$O_2^- + O_2 + M \longrightarrow O_2^- \cdot O_2 + M \quad (55)$$

$$O_2^- + N_2 + M \longrightarrow O_2^- \cdot N_2 + M \quad (56)$$

The rate coefficients for these reactions, for which the third body, M, was helium, have been determined at 200 K in a FA by Adams et al.[76] and very recently their temperature dependence has been studied by Fehsenfeld[182] who finds that the rate coefficients vary as $T^{-2.5}$ [reaction (55)] and $T^{-4.2}$ [reaction (56)]. It is clearly very important in atmospheric modelling to appreciate that these reactions can have such rapid temperature dependences. A body of evidence exists which indicates that the magnitudes of the rate coefficients for ternary association reactions increases by about a factor of two when N_2 rather than He is used as the stabilising third body, although little is known concerning the influence of different third bodies on the temperature dependences. Nevertheless, simple rate calculations show that both reactions (55) and (56) are probably the dominant loss processes for O_2^- in all of the atmospheric regions, with reaction (56) being relatively more important at the lowest temperatures. A competing loss process for O_2^- in the stratosphere is the binary charge transfer reaction[183]

$$O_2^- + O_3 \longrightarrow O_3^- + O_2 \tag{57}$$

although it is never dominant.

Following reactions (55) and (56), the weakly bonded ions $O_2^- \cdot O_2$ and $O_2^- \cdot N_2$ undergo fast binary switching reactions with other abundant atmospheric gases: e.g.

$$O_2^- \cdot O_2 + CO_2 \longrightarrow O_2^- \cdot CO_2 + O_2 \tag{58}$$

In this reaction, the more strongly bonded $O_2^- \cdot CO_2$ ion is formed which then reacts with the relatively less abundant NO:

$$O_2^- \cdot CO_2 + NO \longrightarrow O_2^- \cdot NO + CO_2 \tag{59}$$

A detailed study of these switching reactions has been made by Adams et al.[76] who determined the relative bond strengths of a series of molecules to both O_2^- and O_2^+ and recognised that the $O_2^- \cdot NO$ ion formed in reaction (59) is not the stable nitrate ion, NO_3^-, but the more reactive peroxy isomer which reacts with NO:

$$O_2^- \cdot NO + NO \longrightarrow NO_2^- + NO_2 \tag{60}$$

In the D-region and the stratosphere, the relatively stable NO_2^- so produced will react with O_3 producing the nitrate ion[184]

$$NO_2^- + O_3 \longrightarrow NO_3^- + O_2 \tag{61}$$

The extraordinary stability of the NO_3^- ion against electron loss is emphasised by the very recent re-determination of the electron affinity of NO_3 (EA = 4.01 eV)[185] which indicates a value even higher than previously thought.

So the above ternary association reactions and binary switching reactions convert the O_2^- ions to the stable NO_2^- and NO_3^- ions. Other parallel (but probably less significant) reaction schemes which have been proposed involve the production of HCO_3^- and CO_3^- (e.g. from the O_3^- and CO_2 reaction[186]). Significantly, however, CO_3^- reacts with NO and NO_2 to produce NO_2^- and NO_3^- respectively.

Ternary association reactions of the NO_3^- (and the NO_2^-, CO_3^- and HCO_3^-) ions with H_2O can then proceed, thus producing the observed cluster ions

$$\text{e.g.}\quad NO_3^- + H_2O + N_2 \longrightarrow NO_3^- \cdot H_2O + N_2 \tag{62}$$

A good deal of laboratory data have been obtained recently for the rate coefficients and equilibrium constants for the ternary association reactions of several atmosphere negative ions with H_2O[94, 187–191].

The $NO_3^- \cdot (H_2O)_n$ clusters are expected to be relatively unreactive in the atmosphere except in regions where appreciable concentrations of strong acids exist. It is known[119, 156] that HNO_3 will replace H_2O in these clusters and the product ions can then undergo further clustering with HNO_3

$$NO_3^- \cdot H_2O \xrightarrow{HNO_3} NO_3^- \cdot HNO_3 \xrightarrow[M]{HNO_3} NO_3^- (HNO_3)_2 \text{ etc.} \tag{63}$$

So ions of the type NO_3^- (acid)$_n$ could well exist in the stratosphere. However it has recently been shown[219] that NO_3^- clusters react with H_2SO_4 replacing the NO_3^- core ion with HSO_4^-, e.g.

$$NO_3^- \cdot HNO_3 + H_2SO_4 \longrightarrow HSO_4^- \cdot HNO_3 + HNO_3 \tag{64}$$

Such replacement reactions would be most important in the troposphere. However, the large relative concentrations of water vapour in this region would tend to shift the equilibrium in the generalised reaction

$$HSO_4^- (\text{acid})_m + H_2O \rightleftharpoons HSO_4^- (H_2O)_n + \text{acid} \tag{65}$$

to the right, thus returning the HSO_4^- hydrates as the dominant species. The minor neutral stratospheric and tropospheric species such as N_2O, N_2O_5 and the freons, will surely play a part in the overall negative ion chemistry[163], but much more laboratory data and in-situ observations are necessary before this can be quantitatively assessed.

3.2.4 Electron-Ion Dissociative Recombination

When an electron neutralizes a positive ion, the energy released can be dissipated either in photon emission (radiative recombination), or by a third body encounter with the transient excited atom or molecule (three-body recombination) or by the fragmentation of the transient excited molecule (dissociative recombination). Radiative recombination only occurs with a very small probability and three-body recombination only occurs at high pressures or high charge densities, neither of these being appropriate to the atmospheric plasma. It is the dissociative process, exemplified by reactions (5a) and (5b), which is dominant in the ionosphere. In fact, reactions (5a) and (5b) are almost entirely responsible for the loss of ionization in the ionosphere above 85 km altitude (with N_2^+ recombination contributing somewhat) as is readily shown by simple calculations based on laboratory determinations of dissociative recombination coefficients, α_{re}, for the dominant molecular ions O_2^+ and NO^+.

Several laboratory groups have contributed to the determination of dissociative recombination coefficients during the last decade or so, but the greatest contribution has been made by Biondi and his co-workers[102, 104, 192–196]. A comprehensive review of the experimental and theoretical work up to 1970 has been given by Bardsley and Biondi[103] but much laboratory work of significance to the ionospheric plasma has been carried out since, including the determination of α_{re} for cluster ions (see below).

The several independent measurements of α_{re} (O_2^+) made using different techniques collectively represent a most satisfactory situation. The several determinations of α_{re} (O_2^+) at 300 K achieved using the SA-microwave technique[19, 102, 193], the SA-Langmuir probe technique[99, 101] and the value inferred at 300 K from a shock-tube experiment[198] are all in excellent agreement (α_{re} (O_2^+) at 300 K = 2×10^{-7} cm^3 s^{-1}). Also the dependences of α_{re} (O_2^+) on the electron temperature, T_e, determined in the SA, in the shock tube and in the ion trap experiment of Walls and Dunn[199] are quite consistent over a wide range of temperature (encompassing thermospheric temperatures) and establish that α_{re} $(O_2^+) \sim T_e^{-0.6}$. That the shock tube data (taken under conditions of equal electron and ion temperature) are in such good agreement with the SA and ion trap data (taken under conditions of elevated electron temperature only) implies that α_{re} (O_2^+) is independent of the ion temperature over the temperature range of the shock tube experiment (600 K – 2500 K). To complete this satisfying picture, Torr et al.[200] have deduced a dependence of α_{re} (O_2^+) on T_e from AE-C satellite data which apparently is not inconsistent with the laboratory data.

For α_{re} (NO^+), however, the situation is not so clear. The 300 K value is well established (α_{re} $(NO^+) = 4.3 \times 10^{-7}$ cm^3 s^{-1})[192, 201], but serious disagreement exists between the T_e dependence of α_{re} (NO^+) obtained from the SA experiment of Huang et al.[195] and the ion trap experiment of Walls and Dunn[199] which give α_{re} $(NO^+) \sim T_e^{-0.37}$ and $\sim T_e^{-0.83}$ respectively. The recently reported T_e dependence by Torr et al.[202] deduced from the night-time AE satellite observations strongly favours that obtained using the ion trap experiment. In a very recent review paper, Torr and Torr[36] reason that the long trapping times for the NO^+ ions in the ion trap experiment allows the ions to relax to the ground vibrational state as is also expected for NO^+ ions in the night-time thermosphere, whereas they note that Huang et al. have indicated that the NO^+ ions in their afterglow experiment are probably somewhat vibrationally excited. Hence the good agreement between the ion trap and satellite data. However, Huang et al. have carefully considered from the standpoint of theory the likely influence on the dissociative recombination process of vibrational excitation in the NO^+ ions and conclude that it cannot account for the slower decrease of α_{re} (NO^+) with T_e that is observed in the SA experiment. Thus, the problem remains unresolved but does serve to emphasise that the population distributions of internal states of ions in the best conceived laboratory experiments and in the ionosphere may often be very different and that laboratory data should be applied to the ionospheric plasma with great care.

Determination of the dissociative recombination coefficients of cluster ions is very challenging experimentally since it is almost impossible to establish one cluster ion species as the only positive ion in a plasma and so the raw data has to be deconvoluted to obtain α_{re} for each of the individual ion species which are present. The only significant amount of data obtained to date has again been obtained using the

SA-microwave technique[104, 194, 196] and relates to the $H_3O^+ (H_2O)_n$ and $NH_4^+ (NH_3)_m$ cluster ion series, although data relating to the reaction:

$$O_2^+ \cdot O_2 + e \longrightarrow \text{products} \tag{66}$$

has previously been obtained[101, 102] which indicated that α_{re} $(O_2^+ \cdot O_2)$ is about an order of magnitude greater than α_{re} (O_2^+). The data of Leu et al.[194] for the cluster ion series $H_3O^+ (H_2O)_n$ for n = 0 to 6 showed that α_{re} markedly increased with n and approached the very high value of 10^{-5} cm^3 s^{-1} for n = 6, almost two orders of magnitude greater than α_{re} (O_2^+) at 300 K. This considerably enhanced α_{re} for cluster ions largely explains the rapid reduction in electron density in the ionosphere near the mesopause[203], since it is below that altitude in the D-region (Fig. 3) that significant concentrations of cluster ions begin to be created. Subsequently, Huang et al.[104, 196] have investigated the electron temperature dependence of α_{re} for the cluster ions $NH_4^+ \cdot (NH_3)_{1,2}$ and $H_3O^+ (H_2O)_{1,2,3}$ and found α_{re} to be essentially independent of T_e. This feature of cluster ion dissociative recombination reactions simplifies somewhat the quantitative modelling of the ion chemistry of the D-region.

As is clear from Sect. 3.2.3, several species of cluster ions will be simultaneously present in the mesosphere and so ideally for detailed de-ionization rate calculations, the dissociation recombination coefficients are needed for them all including, for example, the $NO^+ \cdot (H_2O)_n$ ions. Such data will be difficult to obtain and in the meantime estimates of α_{re} have to be relied upon. Based on the available data, the very least that can be said is that α_{re} for all cluster ion reactions will be relatively large ($\gtrsim 10^{-6}$ cm^3 s^{-1}).

A desirable, but difficult extension of dissociative recombination studies would be the determination of α_{re} for excited ions such as vibrationally excited ions in their ground electronic states and the electronically excited $O_2^+(^4\Pi_u)$ and NO^+ $(^3\Sigma)$ ions all of which undoubtedly exist in the ionosphere. As well as being of ionospheric significance, such data would also be of fundamental interest in that it would assist in clarifying theoretical models of the dissociative recombination process.

3.2.5 Ion-Ion (Ionic) Recombination

Ionic recombination is a general term used to describe the charge neutralization processes which can occur when a positive ion and a negative ion interact producing neutral particles. Three distinct mechanisms have been characterised according to how the energy released in the interaction is dispersed, as in the case of electron-ion recombination. They are

(i) radiative recombination (photon emission)

(ii) mutual neutralization (energy converted into internal and/or kinetic energy of the product neutrals) and

(iii) ternary or 3-body recombination (energy carried away largely as internal and/or kinetic energy of a neutral third body).

Radiative ionic recombination only occurs with a very small probability and is unimportant in the ionosphere and so will not be discussed further. Mutual neutralization, exemplified by reaction (9) and the simpler case:

$$NO^+ + NO_2^- \longrightarrow (NO)^* + (NO_2)^* \tag{67}$$

and ternary recombination, exemplified by

$$NO^+ + NO_2^- + N_2 \longrightarrow NO + NO_2 + (N_2)^* \tag{68}$$

where the asterisks denotes internal and/or kinetic excitation, are both important in the lower atmosphere.

Both experimental and theoretical studies of ionic recombination have a long history dating from the beginning of this century (see the recent review by Flannery[204)]). It is sufficient to say here that most of the experimental work carried out prior to 1960 was concerned with studies of the ternary process which dominates the loss of ionization from an electronegative gas at high pressures (above about 0.1 atmosphere pressure). This process is recognised by an increasing efficiency of recombination with increasing background gas pressure and several experiments have established the order of magnitude of the ternary ionic recombination coefficient, including the most recent experiments of Mahan and Person[205)]. However in none of the experiments have the ions been positively identified (by mass spectrometry) and so many of these experiments are of dubious validity, especially the very early experiments for which only poor vacuum techniques were available. The available data on binary and ternary ionic recombination up to 1973 has been reviewed by Mahan[206)].

Notwithstanding the uncertainties in the ternary association data, it is clear that the ternary process will dominate ionic recombination in the lower stratosphere and troposphere where the pressure is high. However, in the upper stratosphere and the D-region the pressure is relatively low and the binary process of mutual neutralization becomes the major ionic recombination loss process. The laboratory data available on this process was very sparse prior to 1970[206)] and the nature of the recombining ions was uncertain. With the advent of the merged beam technique[207, 208)] a concerted effort was made to accumulate data on mutual neutralization, especially for reactions of atmospheric interest. In common with all beam techniques, the cross-section as a function of interaction energy is the determined parameter in the merged beam experiment and the rate coefficient at a given temperature, which is the required parameter in calculations of deionization rates in a plasma, is calculated from the cross-section data. In this manner the rate coefficients for several mutual neutralization reactions have been obtained, including that for reaction (67) and also for the reaction:

$$NO^+ + NO_3^- \longrightarrow \text{products} \tag{69}$$

both reactions (67) and (69) being of interest in D-region studies. The data obtained from the merged beam experiment together with a comprehensive discussion of the theory of the mutual neutralization process is contained in a review by Moseley et al.[209].

Following their successful application of the Langmuir probe technique to the determination of electron and ion densities and electron temperatures in flowing afterglow plasmas[116, 117], Smith and his colleagues have recently used the technique to study mutual neutralization reactions[118–120]. The first reactions studied were (67) and (69) and binary rate coefficients, α_{ri}, were obtained under collision-dominated conditions for which the kinetic temperature of the interacting ions could be defined and in which the rotational-vibrational states of the ions were expected to be their thermal equilibrium distributions. The α_{ri} obtained for both these reactions at 300 K was about one order of magnitude less than those inferred from the merged beam experiment at an equivalent temperature. Significantly, however, the FA measurements are in good agreement with theoretical estimates[118]. These results have cast some doubt on the validity of the merged beam technique for the determination of thermal energy ionic recombination coefficients especially for reactions of molecular ions. It is likely that electronically excited ions may be a significant component of the ion beams but this alone cannot account for the very high values obtained for α_{ri}[209] and the major errors might be introduced in the conversion of the cross-section data to thermal energy rate coefficients. Fewer complications exist in the FA which is an inherently more appropriate medium in which to study thermal energy reactions. A study of the temperature dependence of reaction (67) over the temperature range 185–530 K[210], showed that $\alpha_{ri} \propto T^{-1/2}$ in accordance with theory, giving added credence to the FA results.

Subsequent FA studies[119, 120] have determined α_{ri} for several reactions including some ionospherically important cluster ion reactions, for example:

$$H_3O^+(H_2O)_3 + NO_3^- \longrightarrow \text{products} \tag{70}$$

$$H_3O^+(H_2O)_3 + NO_3^-(HNO_3) \longrightarrow \text{products} \tag{71}$$

An especially important result from these studies is that α_{ri} is remarkably independent of the complexity of the reacting ions (in marked contrast to electron dissociative recombination), only varying over the limited range $(4–10) \times 10^{-8}$ cm^3 s^{-1} at 300 K, even for ions as different as those involved in reactions (67) and (71). This coupled with the relatively weak temperature dependence of α_{ri} in practice allows a single value for α_{ri} ($\sim 6 \times 10^{-8}$ cm^3 s^{-1}) to be used for all mutual neutralization reactions in ionospheric de-ionization calculations without introducing serious errors. This value is in close accordance with estimates of ionic recombination coefficients obtained by Ulwick[211] from observations of ionization production and loss rates in the atmosphere in the altitude region 50–75 km.

The accumulated mutual neutralization data from the FA experiments together with that available for the ternary recombination process, indicates that binary mutual neutralization is the dominant ionic recombination process above about 30 km in the atmosphere whereas below this altitude the ternary process becomes dominant[210]. It should be stressed that this generalization is based on dubious ternary recombination data and indeed on mutual neutralization data for moderate-sized clusters only.

Whilst the FA mutual neutralization data is probably the most reliable obtained to date, further data is required especially for the mutual neutralization of large clusters and there is a great need for reliable data on the ternary recombination process over the whole range of atmospheric pressures. It is also of interest from an atmospheric as well as a fundamental viewpoint to identify the neutral products of these neutralization reactions. This has been achieved for the relatively simple reaction (67)[212], but it will be much more difficult for cluster ion recombination reactions.

As is mentioned in Sect. 2.2, a discussion of de-ionization processes in the Earth's atmosphere would be incomplete without a mention of the rôle of aerosols. The attachment of ions to aerosols in the stratosphere and troposphere has been considered by several workers[213]. It is clear that their presence will enhance the loss of ions from the gas phase at a rate dependent on the nature, size and number density of the particles, and so this process, which could be the dominant ionization loss process, must be considered along with gas phase ionic recombination in detailed atmospheric de-ionization rate calculations.

4 Current Status of Atmospheric Ion Chemistry

The current understanding of the ion chemistry of the atmosphere has been achieved by co-ordinating the data obtained from in-situ ion composition measurements with the data obtained from appropriate laboratory experiments. This review has largely been concerned with the elementary ionic reaction processes involved in the overall chemistry and detailed chemical models of the ion chemistry of the atmosphere have been deliberately excluded since such have recently appeared in the literature[8, 73, 74, 147]. However, it is appropriate here to summarise, through block diagrams, the chains of ionic reactions via which ions are formed, evolve and are finally lost from the atmosphere. To this end, it is convenient to consider separately three regions of the atmosphere:

(i) the thermosphere (essentially the E- and F-regions of the ionosphere) in which the charged particles are positive ions and electrons only.

(ii) the mesosphere (essentially the D-region of the ionosphere) in which the charged particles are positive ions, negative ions and electrons, the relative concentrations of the last two varying with altitude.

(iii) the stratosphere and troposphere, in which the charged particles are positive ions and negative ions (together with charged aerosols).

(i) *The Thermosphere.* The most important reactions are summarised in the simple scheme in Fig. 6. Only a few positive ion-neutral particle reactions are involved, all binary, which convert the primary ions (see caption to Fig. 6) into the terminating molecular ions, these eventually being lost via dissociative recombination with electrons. The thick inter-connecting lines in Fig. 6 indicate channels of enhanced reactivity or additional reaction channels which open when the reactant ions are electronically excited (specifically O^+ (^2D, ^2P) and O_2^+ (a $^4\Pi_u$)). The net effect of the

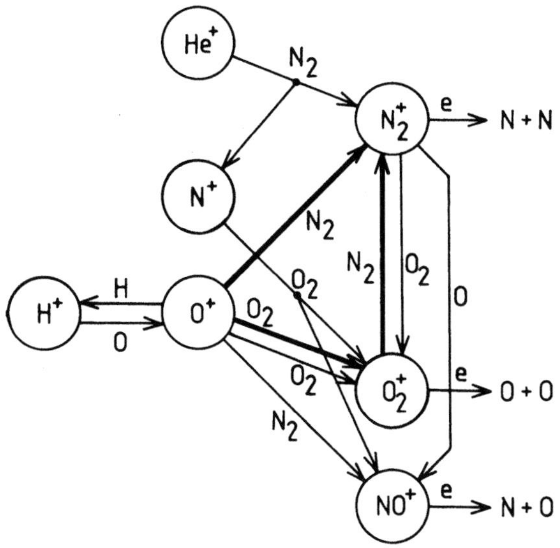

Fig. 6. A limited scheme describing the ion chemistry of the thermospheric plasma. The ionic species are arranged from top to bottom of the diagram in order of reducing recombination energy. The thick interconnecting lines indicate channels of enhanced reactivity or channels which open when the reactant ions are electronically excited. The major primary ions are H^+, He^+, N^+ and O^+ in the upper part of the region and O^+, N_2^+ and O_2^+ in the lower part of the region. The major terminating ions throughout the region are NO^+ and O_2^+

presence of any significant fractions of excited ions, and also of doubly-charged ions, would be to modify the equilibrium proportions of the terminating molecular ions. Similarly, if appreciable fractions of the O_2 and N_2 molecules in the thermosphere are vibrationally excited (as theoretical estimates indicate[214)]), the rate coefficients and branching ratios of the reactions involving these species could be radically changed, especially those reactions which are slow for ground state reactants [e.g. reaction (16)]. The influence of vibrational excitation within either the ions or the molecules on ion-molecule reaction rates and products is little understood and is an area where a real need exists for laboratory data. Also little data exists concerning the influence of internal excitation on the dissociative recombination coefficients of atmospheric ions. Such data, when available, will contribute towards a more detailed understanding of the chemistry of the thermosphere.

(ii) *The Mesosphere.* Generalised reaction schemes for both the positive ion and the negative ion chemistry are presented in Fig. 7. The separate ion chemistries evolve through the chain of binary and ternary reactions A_+ to E_+ and A_- to E_- and, as indicated, positive ions can be lost at any point along the chain by either dissociative recombination or mutual neutralization. Negative ions can be lost by mutual neutralization, associative detachment and photodetachment. Whilst it appears from Fig. 7 that the negative ion chemistry is simpler than that for the positive ions, this is not so but simple indicates a greater degree of ignorance regarding the negative ion chemistry. The initial reactions in both schemes (B_+, C_+ and A_-, B_-, C_-) have been reasonably well studied in the laboratory and it now seems probable that their rôles in mesospheric ion chemistry have been properly identified. However, there still exists a great deal of uncertainty about stages D_+, E_+, D_- and E_-. In particular, the rôle of the mixed cluster ions, e.g., $NO^+ \cdot H_2O \cdot M$, in the evolution of the $H_3O^+(H_2O)_n$ ions is not understood, a problem which will only be resolved by laboratory measurements of the rate coefficients and their temperature dependences for the large number of ternary association reactions expected to be involved. The problems are no less acute regarding the negative ion chemistry. The advent of the temperature-variable SIFT provides an opportunity to clarify some of these problems, although an extensive programme of measurements is required.

Some problems remain regarding the neutralization rates of the ions since although the dissociative recombination coefficients, α_{re}, of the positive ions, including the water cluster ions, have been reasonably well studied estimate have to be made for α_{re} for several important reactions as yet not studied. Similarly, although mutual neutralization rate coefficients, α_{ri}, appear to be largely independent of the nature of the ions even when they are relatively big clusters, α_{ri} should still be determined for the larger clusters now known to exist in the atmosphere.

(iii) *The Stratosphere and Troposphere.* Any discussion of the ion chemistry of these regions must inevitably be more speculative than that for the higher altitude regions, largely due to the sparcity of observational data regarding the ion composition at low altitudes, but also because of the much greater complexity of the neutral atmosphere which can compound the number of reactions occurring. Figure 8 is a general representation of the positive ion and negative ion evolution in the ion-ion plasma and necessarily contains many of the elements included in Fig. 7. Again the initial reactions are well understood and it is the stages D_+, E_+, D_- and E_- which are very uncertain. What is not in doubt is that a great deal more reactions will be involved compared with the corresponding stages of mesospheric ion chemistry, and marked global, altitudinal and local variations will occur. Minor neutral atmospheric constituents are expected to play vital rôles, the oxides of nitrogen (e.g. N_2O, N_2O_5) and acids (e.g., HNO_3, HCl) influencing the negative-ion chemistry and the bases (e.g., NH_3, NaOH) influencing the positive-ion chemistry. The sizes of the charged particles will be wide ranging, varying from those amenable to mass spectrometry to charged droplets or aerosols. Unfortunately the great difficulty in mass spectrometric sampling of ions from such a high pressure, low ionization density plasma has so far prevented identification of the ion types in the troposphere. However existing techniques are continuously being improved and new techniques are being conceived so that such sampling will surely be achieved in the not-too-distant future.

Loss of ions occurs via the processes of mutual neutralization, ternary ionic recombination and attachment to aerosol surfaces, processes which urgently need further study in the laboratory. It is an interesting fact that the ion chemistry directly accelerates the loss of ionization from all regions of the atmospheric plasma. Atomic ions are converted into molecular ions, molecular ions into larger cluster ions which recombine more rapidly. The larger ions also act as nucleation sites for the formation of aerosols, thus involving a transition from the molecular to the liquid state. The aerosols then provide efficient adsorption/recombination surfaces for ions from the gas phase.

Low temperature plasmas, not least the environmental plasma, are indeed interesting and complex media.

Acknowledgment. We are grateful to Dr. Eldron F. Ferguson for his constructive comments on the manuscript.

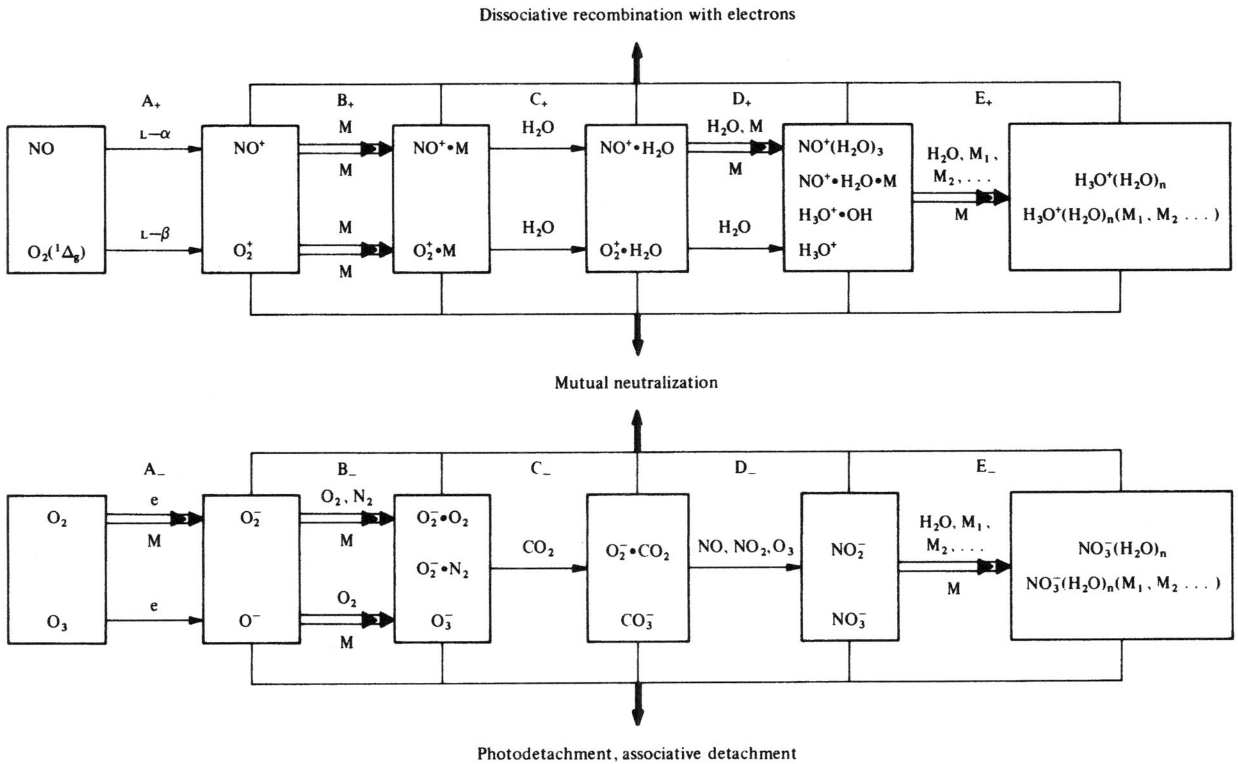

Fig. 7. A generalized scheme describing the ion chemistry of the mesospheric plasma. Binary reactions are indicated by single arrows and ternary reactions by double arrows with M representing the neutral third body (usually N_2 or O_2). M_1 and M_2 represent any major neutral atmospheric constituents. The letters $A_{+,-}$ to $E_{+,-}$ designate the particular reaction steps and are only included to assist in discussion (see text). The positive ion scheme originating from O_2^+ and the negative ion scheme originating from O^- are considered to be of lesser importance in the overall chemistry

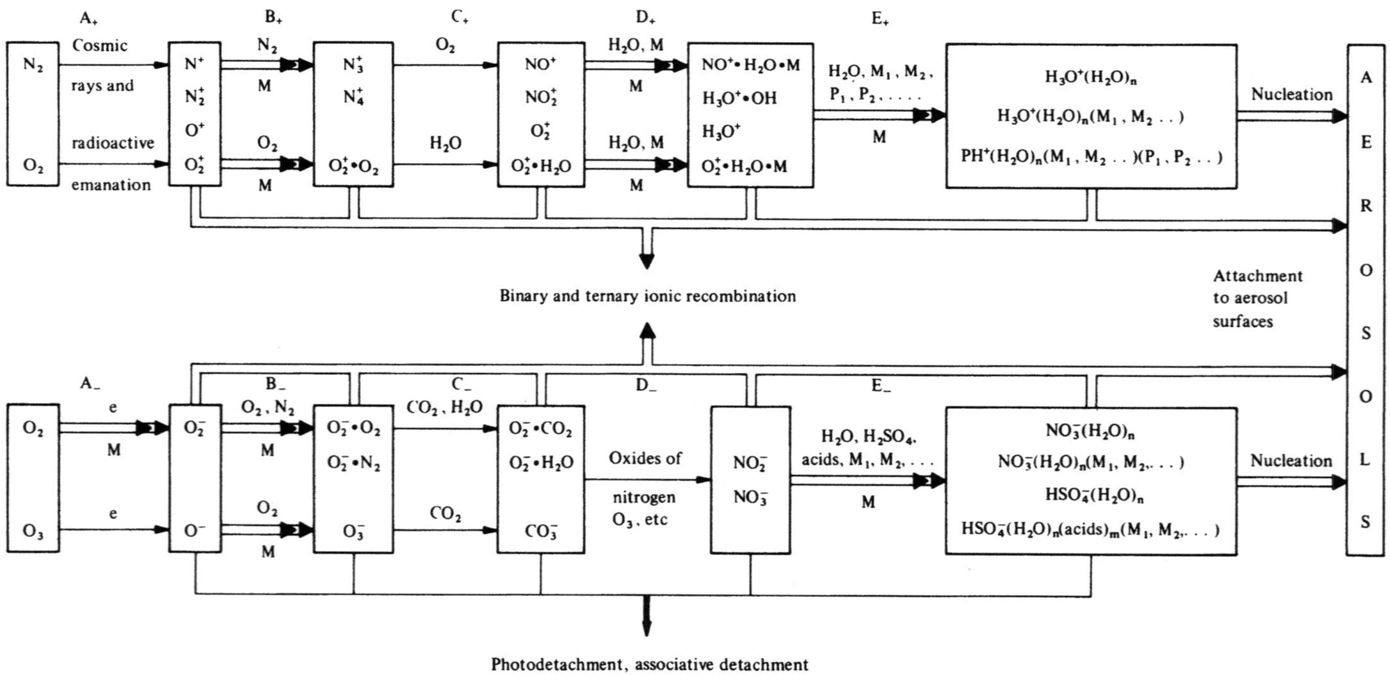

Fig. 8. A generalized scheme describing the ion chemistry of the stratosphere and troposphere. The symbols are as in Fig. 7 but additionally with P representing bases such as NH_3 and NaOH and (acids) representing HNO_3, H_2SO_4, HCl etc. Those parts of the positive ion schemes involving NO^+ and NO_2^+ and the negative ion scheme originating from O^- are considered to be of lesser importance in the overall chemistry

References

1. Ratcliffe, J. A. (ed.): Fifty years of the ionosphere. J. Atmos. Terres. Phys. *36*, 2069 (1974)
2. Ratcliffe, J. A. (ed.): Physics of the upper atmosphere. New York: Academic Press 1960
3. Ratcliffe, J. A.: An introduction to the ionosphere and magnetosphere. Cambridge: Cambridge Univ. Press 1972
4. McEwan, M. J., Phillips, L. F.: Chemistry of the atmosphere. London: Edward Arnold 1975
5. Banks, P. M., Kockards, G.: Aeronomy. New York: Academic Press 1973
6. U. S. Standard Atmosphere: Publication No. NOAA-S/T 76–1562. Washington D.C. 1976
7. Nicolet, M.: Revs. Geophys. Space Phys. *13*, 593 (1975)
8. Thomas, L.: Radio Sci. *9*, 121 (1974)
9. Meira, L. G.: J. Geophys. Res. *76*, 202 (1971)
10. Mitra, A. P.: Endeavour *2*, 12 (1978)
11. Evans, W. F. J., Llewellyn, E. J.: Ann. Geophys. *26*, 167 (1970)
12. Dickinson, P. H. G., Twiddy, N. D., Young, R. A.: Space Research XVI, Berlin: Akademie-Verlag. 301 (1976)
13. Ehhalt, D. H.: Revs. Geophys. Space Res. *16*, 217 (1978)
14. Schmeltekopf, A. L. et al.: J. Atmos. Sci. *34*, 729 (1977)
15. Stewart, R. W., Hoffert, M. I.: J. Atmos. Sci. *32*, 195 (1975)
16. Banks, P. M.: Ann. Geophys. *22*, 577 (1966)
17. Thrush, B. A.: Nature *276*, 345 (1978)
18. Schmeltekopf, A. L., et al.: Geophys. Res. Letts. *2*, 393 (1975)
19. Friedman, H.: The Sun's ionizing radiations. In: Physics of the upper atmosphere. Ratcliffe, J. A. (ed.), pp. 133–218. New York: Academic Press 1960
20. Johnsen, C. Y.: J. Geophys. Res.: *71*, 330 (1966)
21. Narcisi, R. S., Bailey, A. D., Wlodyka, L. E., Philbrick, C. R.: J. Atmos. Terres. Phys. *34*, 647 (1972)
22. Arnold, F., et al.: J. Atmos. Terres. Phys. *33*, 1169 (1971)
23. Cole, R. K. Jr., Pierce, E. T.: J. Geophys. Res. *70*, 2735 (1965)
24. Smith, L. G.: Radio Sci. *1*, 178 (1966)
25. Evans, J. V.: J. Atmos. Terres. Phys. *36*, 2183 (1974)
26. Fehsenfeld, F. C., Ferguson, E. E.: J. Chem. Phys. *56*, 3066 (1972)
27. Ferguson, E. E.: Revs. Geophys. *5*, 305 (1967)
28. Oppenheimer, M., et al.: J. Geophys. Res. *82*, 5485 (1977)
29. Oppenheimer, M., Dalgarno, A., Brinton, H. C.: J. Geophys. Res. *81*, 3762 (1976)
30. Ferguson, E. E.: Ion Chemistry. In: Atmospheres of Earth and the planets. McCormac, B. M. (ed.), pp. 197–210, Dordrecht, Holland: D. Reidel 1975
31. Ferguson, E. E.: Ionospheric ion-molecule reactions. In: Interactions between ions and molecules. Ausloos, P. (ed.), pp. 313–339. New York: Plenum Press 1975
32. Bates, D. R.: J. Atmos. Terres. Phys. *36*, 2287 (1974)
33. Dalgarno, A., McElroy, M. B.: Planet Space Sci. *13*, 947 (1965)
34. Torr, D. G., Orsini, N.: Planet. Space Sci. *25*, 1171 (1977)
35. Orsini, N., et al.: J. Geophys. Res. *82*, 4829 (1977)
36. Torr, D. G., Torr, M. R.: Revs. Geophys. Space Phys. *16*, 327 (1978)
37. Brieg, E. L., et al.: J. Geophys. Res. *82*, 1008 (1977)
38. Lindinger, W., et al.: J. Chem. Phys. *62*, 4101 (1975)
39. Glosik, J., et al.: J. Phys. B *11*, 3365 (1978)
40. Tichy, M., et al.: Int. J. Mass Spectrom. Ion Phys. *29*, 231 (1979)
41. Dotan, I., Albritton, D. L., Fehsenfeld, F. C.: J. Chem. Phys. *71*, 3280 (1979)
42. Spears, K. G., et al.: J. Chem. Phys. *56*, 2562 (1972)
43. Adams, N. G., Smith, D., Grief, D.: J. Phys. B *12*, 791 (1979)
44. Smith, D., Grief, D., Adams, N. G.: Int. J. Mass Spectrom. Ion Phys. *30*, 271 (1979)
45. Johnsen, R., Biondi, M. A.: Geophys. Res. Letts. *5*, 847 (1978)
46. Narcisi, R. S.: Space Research *8*, 360 (1968)

47. Narcisi, R. S.: Ion and neutral composition measurements in the lower ionosphere. In: Methods of measurement and results of lower ionospheric structure. Rawer, K. (ed.), pp. 207–218. Berlin: Akademie-Verlag 1974
48. Goldberg, R. A., Witt, G.: J. Geophys. Res. *82,* 2619 (1977)
49. Swider, W.: Ann. Geophys. *26,* 595 (1970)
50. Ferguson, E. E., Fehsenfeld, F. C.: J. Geophys. Res. *73,* 6215 (1968)
51. Fehsenfeld, F. C.: Can. J. Chem. *47,* 1808 (1969)
52. Kebarle, P.: Higher-order reactions – ion clusters and ion solvation. In: Ion-molecule reactions Vol. 1. Franklin, J. L. (ed.), pp. 315–362. London: Butterworths 1972
53. Keller, G. E., Beyer, R. A.: J. Geophys. Res. *76,* 289 (1971)
54. Johnsen, R., Brown, H. L., Biondi, M. A.: J. Chem. Phys. *55,* 186 (1971)
55. Narcisi, R. S., Bailey, A. D.: J. Geophys. Res. *70,* 3687 (1965)
56. Narcisi, R. S., Roth, W.: Adv. Electron. Electron Phys. *29,* 79 (1970)
57. Goldberg, R. A., Aikin, A. C.: EOS Trans AGU *53,* 1077 (1972)
58. Krankowsky, D., Arnold, F., Wieder, H. In: Magnetosphere-ionosphere interactions. Folkestad, K. (ed.), pp. 19–28. Oslo: Oslo University Press, 1972
59. Arnold, F., Krankowsky, D.: 4th Int. COSPAR Symp. Solar Terres. Phys. 17th June – 1st July, 1974, San Paulo, Brazil
60. Arnold, F., Krankowsky, D., In: Dynamical and chemical coupling. Grandal, B., Holtet, J. A. (eds.), pp. 93. Dordrecht, Holland: D. Riedel. 1977
61. Arnold, F., Krankowsky, D.: J. Atmos. Terres. Phys. *39,* 625 (1977)
62. Arnold, F., Fabian, R.: Nature *283,* 55 (1980)
63. Fehsenfeld, F. C., Ferguson, E. E.: J. Geophys. Res. *74,* 2217 (1969)
64. Lineberger, W. L., Puckett, L. J.: Phys. Rev. *187,* 286 (1969)
65. Good, A., Durden, D. A., Kerbarle, P.: J. Chem. Phys. *52,* 222 (1970)
66. Puckett, L. J., Teague, M. W.: J. Chem. Phys. *54,* 2564 (1971)
67. Fehsenfeld, F. C., Mosesman, M., Ferguson, E. E.: J. Chem. Phys. *55,* 2115, 2120 (1971)
68. Howard, C. J., Rundle, H. W., Kaufman, F.: J. Chem. Phys. *55,* 4772 (1971)
69. French, M. A., Hills, L. P., Kebarle, P.: Can. J. Chem. *51,* 456 (1973)
70. Ferguson, E. E.: Laboratory measurements of D-region ion-molecule reactions. In: Mesospheric models and related experiments. Fioccio, G. (ed.), pp. 188–197. Dordrecht, Holland: D. Reidel 1971
71. Thomas, L.: J. Atmos. Terres. Phys. *38,* 61, 1345 (1976)
72. Reid, G. C.: Planet. Space Sci. *25,* 275 (1977)
73. Ferguson, E. E., In: Kinetics of ion-molecule reactions. Ausloos, P. (ed.) pp. 377–403. New York: Plenum Press 1979
74. Reid, G. C.: Adv. At. Mol. Phys. *63,* 375 (1976)
75. Sechrist, C. F. Jr.: J. Atmos. Terres. Phys. *34,* 1565 (1972)
76. Adams, N. G., et al.: J. Chem. Phys. *52,* 3133 (1970)
77. Castleman, A. W. Jr., In: Physics and chemistry of upper atmospheres. McCormac, B. M. (ed.), pp. 143. Dordrecht, Holland: D. Reidel 1973
78. Arnold, F., Krankowsky, D., Marien, K. H.: Nature *267,* 30 (1977)
79. Arnold, F., Bohringer, H., Henschen, G.: Geophys. Res. Letts. *5,* 653 (1978)
80. Kebarle, P.: 1979 private communication
81. Ferguson, E. E.: Geophys. Res. Letts. *5,* 1035 (1978)
82. Olsen, J. B. et al.: EOS Trans AGU *58,* 1201 (1977)
83. Arijs, E., Ingols, J., Neuejans, D.: Nature *271,* 642 (1978)
84. Crutzen, P. J., In: Physics and chemistry of upper atmospheres. McCormack, B. M. (ed.), pp. 110, Dordrecht, Holland: D. Reidel 1973
85. Mohnen, V. A.: J. Geophys. Res. *75,* 1717 (1970)
86. Mohnen, V. A.: Pure Appl. Geophys. *84,* 141 (1971)
87. McDaniel, E. W., et al.: Ion-molecule reactions. New York: Wiley 1970
88. Huntress, W. T. Jr.: Astrophys. J. Suppl. Series *33,* 495 (1977)
89. Dickinson, P. H. G., Sayers, J.: Proc. Phys. Soc. *76,* 137 (1960)

90. Sayers, J., Smith, D.: In: Proc. IIIrd int. conf. on the physics of electronic and atomic collisions. McDowell, M. R. C. (ed.), pp. 871–876. Amsterdam: North Holland 1964
91. Sayers, J., Smith, D.: Disc. Faraday Soc. *37*, 167 (1964)
92. Copsey, M. J., Smith, D., Sayers, J.: Planet. Space Sci. *14*, 1047 (1966)
93. Lineberger, W. C., Puckett, L. J.: Phys. Rev. *186*, 116 (1969)
94. Puckett, L. J., Lineberger, W. C.: Phys. Rev. *A 1*, 1635 (1970)
95. Smith, D., Goodall, C. V., Copsey, M. J.: J. Phys. B *1*, 660 (1968)
96. Smith, D., Dean, A. G., Adams, N. G.: Z. Physik *253*, 191 (1972)
97. Dean, A. G., Smith, D., Adams, N. G.: J. Phys. B 7, 644 (1974)
98. Smith, D., Dean, A. G., Adams, N. G.: J. Phys. D *7*, 1944 (1974)
99. Smith, D., Goodall, C. V.: Planet. Space Sci. *16*, 1177 (1968)
100. Smith, D., et al.: J. Phys. B *3*, 34 (1970)
101. Plumb, I. C., Smith, D., Adams, N. G.: J. Phys. B *5*, 1762 (1972)
102. Kasner, W. H., Biondi, M. A.: Phys. Rev. *174*, 139 (1968)
103. Bardsley, J. N., Biondi, M. A.: Adv. Atm. Mol. Phys. *6*, 1 (1970)
104. Huang, C-M., et al.: Phys. Rev. *A 18*, 64 (1978)
105. McDaniel, E. W., Mason, E. A.: The mobility and diffusion of ions in gases. New York: Wiley 1973
106. Kebarle, P.: Thermochemical information from the study of ion-equilibria. In: Interactions between ions and molecules. Ausloos, P. (ed.), pp. 459–487. New York: Plenum Press 1975
107. Kebarle, P.: Ann. Revs. Phys. Chem. *28*, 445 (1977)
108. Kebarle, P., Godbole, E. W.: J. Chem. Phys. *39*, 1131 (1963)
109. Cunningham, A. J., Payzant, J. D., Kebarle, P.: J. Amer. Chem. Soc. *94*, 7627 (1972)
110. Fehsenfeld, F. C., Ferguson, E. E., Schmeltekopf, A. L.: J. Chem. Phys. *44*, 3022 (1966)
111. Ferguson, E. E., Fehsenfeld, F. C., Schmeltekopf, A. L.: Adv. Atm. Mol. Phys. *5*, 1 (1969)
112. Dunkin, D. B., Fehsenfeld, F. C., Schmeltekopf, A. L., Ferguson, E. E.: J. Chem. Phys. *49*, 1365 (1968)
113. Lindinger, W., Fehsenfeld, F. C., Schmeltekopf, A. L., Ferguson, E. E.: J. Geophys. Res. *79*, 4753 (1974)
114. Ferguson, E. E.: Atom. Nucl. Data Tables *12*, 159 (1973)
115. Albritton, D. L.: Atom. Nucl. Data Tables *22*, 1 (1978)
116. Smith, D., Adams, N. G., Dean, A. G., Church, M. J.: J. Phys. D *8*, 141 (1975)
117. Smith, D., Dean, A. G.: J. Phys. B *8*, 997 (1975)
118. Smith, D., Church, M. J.: Int. J. Mass Spectrom. Ion Phys. *19*, 185 (1976)
119. Smith, D., Adams, N. G., Church, M. J.: Planet. Space Sci. *24*, 697 (1976)
120. Smith, D., Church, M. J., Miller, T. M.: J. Chem. Phys. *68*, 1224 (1978)
121. Heimerl, J., Johnsen, R., Biondi, M. A.: J. Chem. Phys. *51*, 5041 (1969)
122. Johnsen, R., Biondi, M. A.: J. Chem. Phys. *59*, 3504 (1973)
123. Chen, A., Johnsen, R., Biondi, M. A.: J. Chem. Phys. *69*, 2688 (1978)
124. McFarland, M., et al.: J. Chem. Phys. *59*, 6610 (1973)
125. Albritton, D. L., et al.: J. Chem. Phys. *66*, 410 (1977)
126. Tang, I. N., Castleman, A. W. Jr.: J. Chem. Phys. *57*, 3639 (1972)
127. Tang, I. N., Lian, M. S., Castleman, A. W. Jr.: J. Chem. Phys. *65*, 4022 (1976)
128. Adams, N. G., Smith, D.: Int. J. Mass Spectrom. Ion Phys. *21*, 349 (1976)
129. Adams, N. G., Smith, D.: J. Phys. B *9*, 1439 (1976)
130. Smith, D., Adams, N. G., Grief, D.: J. Atmos. Terres. Phys. *39*, 513 (1977)
131. Smith, D., Adams, N. G.: Chem. Phys. Letts. *54*, 535 (1978)
132. Adams, N. G., et al.: Chem. Phys. Letts. *61*, 608 (1979)
133. Adams, N. G., et al.: Chem. Phys. Letts. *63*, 166 (1979)
134. Smith, D., Adams, N. G.: Recent advances in flow tubes: measurement of ion-molecule rate coefficients and product distributions. In: Gas phase ion chemistry. Vol. 1. Bowers, M. T. (ed.). pp. 1–44. New York: Academic Press 1979
135. Smith, D., Adams, N. G.: Astrophys. J. *217*, 741 (1977)
136. Smith, D., Adams, N. G.: Astrophys. J. *220*, L87 (1978)

137. Ferguson, E. E., et al.: Planet. Space Sci. *12,* 1169 (1964)
138. Lindinger, W., Albritton, D. L., Fehsenfeld, F. C.: J. Chem. Phys. *62,* 4957 (1975)
139. Farragher, A. L.: Trans. Faraday Soc. *66,* 1411 (1970)
140. Govers, T. R., et al.: Chem. Phys. Letts. *26,* 134 (1974)
141. Marx, R., Mauclaire, G., Fenistein, S.: Chem. Phys. Letts. *33,* 357 (1975)
142. Kemper, P. R., Bowers, M. T.: Chem. Phys. Letts. *36,* 183 (1975)
143. Johnsen, R., Macdonald, J. A., Biondi, M. A.: J. Chem. Phys. *66,* 4718 (1977)
144. Smith, D., Fouracre, R. A.: Planet. Space Sci. *16,* 243 (1968)
145. Ferguson, E. E., et al.: J. Chem. Phys. *50,* 5039 (1969)
146. Schmeltekopf, A. L., et al.: Planet. Space Sci. *15,* 401 (1967)
147. Ferguson, E. E., Fehsenfeld, F. C., Albritton, D. L.: Ion-chemistry of the Earths atmosphere, In: Gas phase ion chemistry. Vol. 1. Bowers, M. T. (ed.). pp. 45–82. New York: Academic Press 1979
148. Dalgarno, A., McElroy, M. B.: Planet. Space Sci. *11,* 727 (1963)
149. Dalgarno, A.: Ann. Geophys. *26,* 601 (1970)
150. Rusch, D. W., et al.: J. Geophys. Res. *82,* 719 (1977)
151. Mauclaire, G. et al.: J. Chem. Phys. *70,* 4023 (1979)
152. Smith, D., Adams, N. G., Miller, T. M.: J. Chem. Phys. *69,* 308 (1978)
153. Dotan, I., et al.: J. Geophys. Res. *83,* 4036 (1978)
154. Phelps, A. V.: Can. J. Chem. *47,* 1783 (1969)
155. Christophorou, L. G.: Atomic and molecular radiation physics. pp. 465. London: Wiley-Interscience 1971
156. Fehsenfeld, F. C., Howard, C. J., Schmeltekopf, A. L.: J. Chem. Phys. *63,* 2835 (1975)
157. Fehsenfeld, F. C.: Associative detachment. In: Interactions between ions and molecules. Ausloos, P. (ed.), pp. 387–412. New York: Plenum Press 1975
158. Hotop, H., Patterson, T. A., Lineberger, W. C.: Phys. Rev. *A 8,* 762 (1973)
159. Fehsenfeld, F. C., Ferguson, E. E.: J. Chem. Phys. *59,* 6272 (1973)
160. Dotan, I., et al.: J. Chem. Phys. *67,* 2874 (1977)
161. Fehsenfeld, F. C., et al.: J. Geophys. Res. *83,* 1333 (1978)
162. Smith, D., Adams, N. G., Henchman, M. J.: J. Chem. Phys. (in press 1980)
163. Davidson, J. A., et al.: J. Chem. Phys. *68,* 2085 (1978)
164. Ferguson, E. E.: Ann. Geophys. *26,* 589 (1970)
165. Reid, G. C.: J. Geophys. Res. *75,* 2551 (1970)
166. Donahue, T.: Radio Sci. *7,* 113 (1972)
167. Dunkin, D. B., et al.: J. Chem. Phys. *54,* 3817 (1971)
168. Ferguson, E. E.: Revs. Geophys. Space Phys. *9,* 997 (1971)
169. Heimerl, J. M., Vanderhoff, J. V.: EOS Trans. AGU *52,* 870 (1971)
170. Niles, F. E., Heimerl, J. M., Keller, G. E.: EOS Trans. AGU *53,* 456 (1972)
171. Heimerl, J. M., Vanderhoff, J. V.: J. Chem. Phys. *60,* 4362 (1974)
172. Johnsen, R., Huang, C-M., Biondi, M. A.: J. Chem. Phys. *63,* 3374 (1975)
173. Arnold, F., Krankowsky, D.: J. Atmos. Terres. Phys. *39,* 625 (1977)
174. Ferguson, E. E., Fehsenfeld, F. C.: J. Geophys. Res. *74,* 5743 (1969)
175. Fehsenfeld, F. C., Mosesman, M., Ferguson, E. E.: J. Chem. Phys. *55,* 2115 (1971)
176. Bohme, D. K., Fehsenfeld, F. C.: Can. J. Chem. *14,* 2715 (1969)
177. Payzant, J. D., Cunningham, A. J., Kebarle, P.: Can. J. Chem. *51,* 3242 (1973)
178. Castleman, A. W. Jr.: Chem. Phys. Letts. *53,* 560 (1978)
179. Castleman, A. W. Jr., In: Kinetics of ion-molecule reactions. Ausloos, P. (ed.), pp. 295–321. New York: Plenum Press 1979
180. Castleman, A. W. Jr.: Space Sci. Revs. *15,* 547 (1974)
181. Good, A.: Chem. Revs. *75,* 561 (1975)
182. Fehsenfeld, F. C.: referred to in Ref.[73)]
183. Fehsenfeld, F. C., et al.: Planet. Space Sci. *15,* 373 (1967)
184. Fehsenfeld, F. C., Ferguson, E. E.: Planet. Space Sci. *16,* 701 (1968)
185. Referred to in Ref.[73)]
186. Parkes, D. A.: J. Chem. Soc. Faraday Trans. I *68,* 627 (1972)

187. Arshadi, M., Kebarle, P.: J. Phys. Chem. *74,* 1483 (1970)
188. Fehsenfeld, F. C., Ferguson, E. E.: J. Chem. Phys. *61,* 3181 (1974)
189. Payzant, J. D., Yamdagni, R., Kebarle, P.: Can. J. Chem. *49,* 3309 (1971)
190. Keesee, R. G., Lee, N., Castleman, A. W. Jr.: J. Am. Chem. Soc. *101,* 2599 (1979)
191. Keesee, R. G., Lee, N., Castleman, A. W. Jr.: J. Geophys. Res. *84,* 3719 (1979)
192. Weller, C. S., Biondi, M. A.: Phys. Rev. *172,* 198 (1968)
193. Mehr, F. J., Biondi, M. A.: Phys. Rev. *181,* 264 (1969)
194. Leu, M. T., Biondi, M. A., Johnsen, R.: Phys. Rev. *A 7,* 292 (1973)
195. Huang, C-M., Biondi, M. A., Johnsen, R.: Phys. Rev. *A 11,* 901 (1975)
196. Huang, C-M., Biondi, M. A., Johnsen, R.: Phys. Rev. *A 14,* 984 (1976)
197. Mentzoni, M. H.: J. Appl. Phys. *36,* 57 (1965)
198. Cunningham, A. J., Hobson, R. M.: J. Phys. *B5,* 2320 (1972)
199. Walls, F. L., Dunn, G. H.: J. Geophys. Res. *79,* 1911 (1974)
200. Torr, D. G., et al.: J. Geophys. Res. *81,* 5578 (1976)
201. Gunton, R. C., Shaw, T. M.: Phys. Rev. *A 140,* 756 (1965)
202. Torr, M. R., St. Maurice, J. P., Torr, D. G.: J. Geophys. Res. *22,* 3287 (1977)
203. Sechrist, C. F. Jr.: Radio Sci. *5,* 663 (1970)
204. Flannery, M. R.: Ionic recombination. In: Atomic processes and applications. Burke, P. G., Moiseiwitsch, B. L. (eds.), pp. 408–466. Amsterdam: North Holland 1976
205. Mahan, B. H., Person, J. C.: J. Chem. Phys. *40,* 392 (1964)
206. Mahan, B. H.: Adv. Chem. Phys. *23,* 1 (1973)
207. Aberth, W., Peterson, J. R., Lorentz, D. C., Cook, C. J.: Phys. Rev. Letts. *20,* 979 (1968)
208. Aberth, W., Peterson, J. R.: Phys. Rev. *A 1,* 158 (1970)
209. Moseley, J. T., Olson, R. E., Peterson, J. R.: Ion-ion mutual neutralization. In: Case studies in atomic physics. McDowell, M. R. C., McDaniel, E. W. (eds.). *5,* 1 (1975)
210. Smith, D., Church, M. J.: J. Phys. B *11,* 4041 (1978)
211. Ulwick, J. C.: Proc. COSPAR symp. on solar particle event of November, 1969. Ulwick, J. C. (ed.) AFCRL-72, 0474 Special Report No. 144,571, Air Force Cambridge Res. Labs. Bedford, MA, 1972
212. Smith, D., Adams, N. G., Church, M. J.: J. Phys. B *11,* 4041 (1978)
213. Hake, R. D. Jr., Pierce, E. T., Viezee, W.: Stratospheric electricity. Final Report, SRI Project 1724, Stanford Research Institute, Menlo Park, Calif. 1973
214. Walker, J. C. G.: In: Physics and chemistry of upper atmospheres. McCormac, B. M. (ed.), pp. 204. Dordrecht, Holland: D. Reidel 1973
215. Smith, D., Adams, N. G.: (unpublished data)
216. Albritton, D. L., et al.: (priv. comm.)
217. Johnsen, R., Biondi, M. A.: (priv. comm.)
218. Victor, G. A., Constantinides, E. R.: Geophys. Res. Letts. *6,* 519 (1979)
219. Viggiano, A. A. et al.: J. Geophys. Res. (in press 1980)

Received April 9, 1979

Plasma-Materials Interactions and Impurity Control in Magnetically Confined Thermonuclear Fusion Machines*

Dieter M. Gruen[1], Stanislav Veprek[2], and Randy B. Wright[1]

1 Chemistry Division, Argonne National Laboratory, Argonne, Illinois 60439, USA
2 Institute of Inorganic Chemistry, University of Zurich, 8057 Zurich, Switzerland

Progress achieved in plasma heating and magnetic confinement during the past decade has brought to the fore a number of problems which have to be solved if controlled thermonuclear fusion is to become an economically and environmentally acceptable energy source. Among them, the interactions of plasmas with solid surfaces represent a very serious obstacle to the achievement of a positive energy balance in the next generation of magnetic fusion devices. In present devices, plasma energy losses are dominated by radiation due to impurities released from the limiter and the vacuum vessel wall. The processes leading to impurity release are complex and still poorly understood, as are the mechanisms of impurity transport. The present chapter summarizes the status of this field at the beginning of 1979. The various elementary processes are discussed and an attempt is made to point out problems towards which future research has to be directed. Various approaches suggested up to now for impurity control as well as potential solutions to the materials erosion problem are discussed. Clearly, the elucidation of plasma-materials interactions presents a challenge to a broad spectrum of chemists, physicists and materials scientists.

Table of Contents

* Work performed in part under the auspices of the US Department of Energy and the Swiss National Science Foundation

1 Introduction

During the last decade the field of controlled thermonuclear fusion has been characterized by a transition from an emphasis on plasma physics experiments toward broad, complex studies with increasing roles being played by engineering, material science, surface science and related disciplines. This change was due to increased understanding of plasma physics resulting in the building and operation of larger and more sophisticated devices with plasmas at higher temperatures and longer confinement times. In turn, a new set of problems arose among which plasma-materials interactions giving rise to plasma impurities have become of major concern.

It has become abundantly clear that the achievement of controlled thermonuclear fusion as an economical and environmentally acceptable energy source represents perhaps the greatest challenge to modern science and technology in that a whole range of complex problems need to be addressed among which plasma impurities is only one, albeit a very important category. The original enthusiasm for nuclear fusion as a "clean, infinite and cheap" energy source has been replaced, after more than twenty five years of research, by a sober view of the magnitude of a wide spectrum of problems that must be solved in order for fusion to become a reality. It took more than thirty years for fission reactors to become commercially available after the demonstration of scientific feasibility and one can hardly expect commercialization to be reached by controlled fusion until early in the next century[1, 2)].

Nevertheless, the need for new energy sources is a strong motivation in the search for future alternatives to the present energy supplies. The significant progress which has been achieved in fusion research during the past few years supports the expectation that scientific feasibility and energy breakeven can be reached by the next generation of large devices such as TFTR (Tokamak Fusion Test Reactor) and JET (Joint European Torus) coming into operation between ~1981 and 1985. These machines should open the way for an experimental fusion power reactor and ultimately a commercial fusion power plant.

Detailed discriptions of the design and engineering requirements for full scale fusion reactors are contained in several excellent reviews which should be consulted for further information[3–5)]. Among the most recent of these is the FRD-II-report[2)] and a review by McCracken and Scott[208)].

This chapter restricts itself to the set of problems associated with plasma materials interaction which in some aspects, are related to plasma chemistry in a broader sense.

2 The Physics and Technology of Controlled Thermonuclear Fusion

First, a brief summary of the physical and technological aspects of the various approaches towards controlled thermonuclear fusion will be given. The emphasis will be on Tokamaks which today present the most promising and best understood approach to fusion energy. Consequently, the following discussion of plasma-materials interactions will be relevant predominantly to Tokamaks. There is, however, no doubt that similar problems arise in differing degrees with other magnetic and inertial confinement systems.

2.1 Thermonuclear Fusion Reactions

The most promising fusion reactions require as fuel D, T or ^3He:

$$D + T \longrightarrow {}^4He\,(3.5) + n\,(14.1) \tag{1}$$

$$D + D \longrightarrow {}^3He\,(0.82) + n\,(2.45) \tag{2}$$

$$\longrightarrow T\,(1.01) + H\,(3.02) \tag{3}$$

$$D + {}^3He \longrightarrow {}^4He\,(3.6) + H\,(14.7) \tag{4}$$

The numbers in parentheses give the energy in MeV carried by the reaction products. Reactions (2) and (3) proceed with comparable probabilities. The disadvantages of reaction (1) are twofold:

1) tritium with a radioactive life-time of 12.5 y is not available in nature and has to be bred and adequately handled in the reactor;

2) the largest part of the energy is carried by the neutrons which have to be thermalized in a blanket surrounding the reactor vacuum vessel.

The blanket contains lithium in order to breed the tritium necessary for continuous long term operation:

$$^6Li + n \rightarrow T + {}^4He \qquad (4.8) \tag{5}$$

$$^7Li + n \rightarrow T + {}^4He + n \qquad (-2.47). \tag{6}$$

One notices that 2.47 MeV is consumed in reaction (6). A significant advantage of this so-called D-T-Li cycle is the possibility of breeding tritium even with natural lithium[3,5,6]. Radiation damage as well as induced radioactivity in structural materials result from the large flux of fast neutrons. Since ~80% of the energy released in the D-T reaction appears as kinetic energy of the neutrons electricity will be chiefly produced via a thermal cycle.

The D-D reactions (2) and (3) do not need fuel breeding since deuterium is a stable isotope available in nature. The ^3He and tritium products can enter the reactions (4) and (1) respectively, giving the so called "D-D-T-^3He" or "catalyzed D-D" cycle[3]. About 40% of the energy yield appears as neutron kinetic energy, the remaining ~40% is carried by charged particles and could be utilized for direct conversion to electricity and 20% appears as radiation. Only charged particles are produced in the D-^3He reaction. Higher-Z materials such as lithium, beryllium and boron, if used as fusion fuels, would provide fuel cycles almost free of tritium and neutrons but would require much higher reaction temperatures. However, the technological requirements for handling of the radioactive tritium, radiation damage, and induced radioactivity by the neutrons might be significantly relaxed if such "advanced" fuel cycles could be made to work in practical systems[7].

A factor in favor of the D-T-Li cycle is its plasma fusion power density (Eq. 9), which reaches, at a temperature of ~10 keV[1], a maximum value which is almost two orders of magnitude higher than that of the D-D-T-^3He cycle at ~60 keV (see Fig. 1).

1 The temperature of thermonuclear plasmas is conveniently described in terms of kiloelectron-volts; 1 keV corresponding to a temperature of 1.1605×10^7 °C

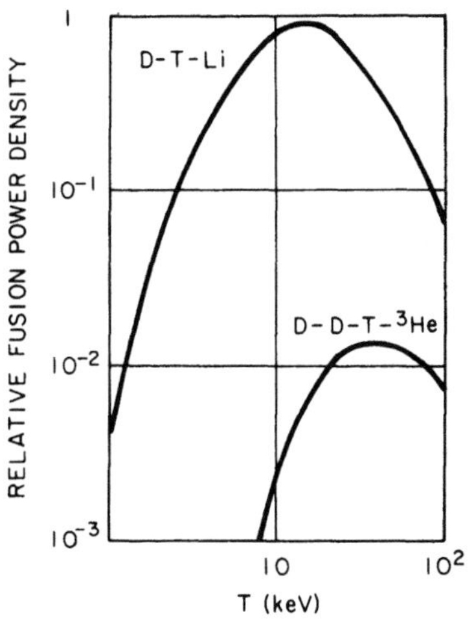

Fig. 1. Relative fusion power densities for the D-T-Li and D-D-T-^{3}He fuel cycles as a function of plasma temperature[3)]

Thus, the D-T-Li fuel cycle is, in spite of its drawbacks, the most likely one to be used in the first fusion reactors. It fulfills the requirement of being an "inexhaustible" energy source since the available lithium resources which represent the limiting factor should be sufficient to meet the future world energy demands for thousands of years[5)].

In order to achieve an overall positive energy balance the D-T plasma has to be heated to the ignition temperature of ~ 5 keV and confined for a sufficiently long time[4)]. Energy breakeven is usually expressed in terms of the Lawson criterion which states that the product of the plasma density, n, and the energy confinement time τ_e has to exceed a critical value which is a function of the plasma temperature[4, 8)]. At a temperature of ~ 10 keV the Lawson criterion requires

$$n \cdot \tau_e \gtrsim 10^{20}\ (m^{-3} s). \tag{7}$$

By the end of 1977 Tokamak devices had achieved $n\tau_e$ values in excess of 10^{19} m^{-3} s and ion temperatures of 2–3 keV. In the early summer of 1978, temperatures in excess of $\gtrsim$ 7 keV were achieved in the Princeton large torus, PLT, using auxilary neutral beam heating[9)]. Other confinement systems such as magnetic mirror devices and intertial confinement are also operating currently below the Lawson criterion but there has been continued development and substantial progress is continually being made.

2.2 Confinement Systems

In this section a brief review is given of the various confinement systems and the state of the art achieved with each one up to the spring of 1979. Readers interested in more details will find them in several excellent reviews: Steiner[3)] and Ribe[4)] summarize the development until 1974; Ref.[10)] gives a review of the world's major facilities for controlled fusion research in 1976 and Ref.[2] is a critical survey of the progress in this field from 1974 to 1977. Further references of interest are by Moss[11] and Bickerton[12].

Two fundamentally different plasma confinement systems are being studied: *magnetic and inertial confinement.* The former utilizes the fact that a charged particle moving in a magnetic field is trapped in a helical trajectory with axes approximately following the magnetic field lines. Confinement of the fusion plasma requires a magnetic field pressure which restrains the kinetic pressure of the hot plasma. Depending on the geometrical configuration, magnetic confinement systems can be divided into two classes called "closed" and "open".

In a closed configuration, the magnetic field lines do not leave the confinement region. A typical example is the toroidal geometry represented by the Tokamak as discussed below. The charged plasma particles can escape from the confining field region due to collisional assisted diffusion across magnetic field lines, to charge exchange processes and to plasma turbulence and other plasma instabilities.

The "open" systems are represented by the *mirror machines* (Fig. 2). Charged particles are trapped in helical orbits along the magnetic field lines between two high field zones, the "magnetic mirrors", in the vicinity of the coils (Fig. 2a). There is a critical angle between the axes of the helical orbit and the axes of symmetry of the magnetic field. Particles moving in an orbit with a smaller angle than the critical one are not reflected by the magnetic mirrors but may escape out of the confinement region. Two consequences are obvious:

1) the energy of the escaping electrically charged particles might be directly converted into electricity;

2) the mirrors operate in a driven mode, i.e. the energy required to heat and sustain the plasma has to be provided by an auxiliary source such as a neutral beam injector.

The simple mirrors (Fig. 2a) are not stable with respect to magnetohydrodynamic instabilities and, therefore, other magnetic field configurations provided by Yin-Yang Coils (Fig. 2b) are used. A representative device using this approach is the 2XIIB-experiment at Livermore[13, 15]. The $n\tau_e$ values obtained in the 2XIIB mirror system, $n\tau_e = 1 \times 10^{17}\ m^{-3}s$, $\bar{E}_i = 13$ keV are far below the requirements for

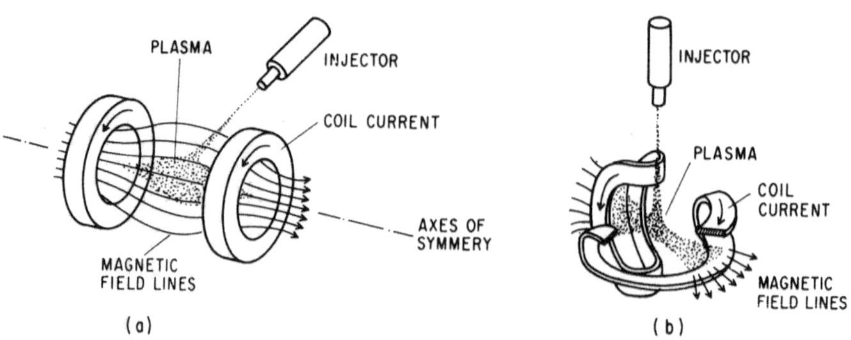

Fig. 2a and b. Magnetic mirror confinement systems: **a** magnetic mirror; **b** Yin-Yang Coils, minimum B magnetic mirror

demonstration of scientific feasibility. The key technological and economical problems with mirror reactors appear to be the availability of adequate neutral beam injectors, direct conversion systems, and the low amplification factor of the plasma which is inherent to the mirror concept[2, 3]. Design studies have begun on mirrors as fusion-fission hybrids[2, 3]. In addition, novel approaches have been considered recently such as the Tandem Mirror and Field Reversed Mirror[14, 15] which appear to offer some advantages.

Several systems, employing both open and closed magnetic confinement such as the Z- and θ-pinches, and the stellarators will now be briefly discussed.

The stellarators are similar to Tokamaks, both employing a closed toroidal magnetic field for basic plasma confinement. The stabilization in stellarators is, however, provided by special helical windings[16]. They possess some potential advantages as compared to the Tokamaks with respect to confinement properties and the possibility of steady state operation. Technical and technological limitations such as the small size of the presently available stellarators and the lack of appropriate, sufficiently intense heating systems have inhibited progress comparable to that which Tokamaks have so far achieved.

The theta-pinch employs a pulsed magnetic field for plasma heating and confinement. The principle of operation is apparent from Fig. 3. After a weak, cold plasma has been formed by means of an auxiliary preionization system, switch S is closed leading to discharging of a capacitor bank C through the coil. The large primary current which flows in the toroidal, i.e. in the "theta" direction induces an axial magnetic field, B, which drives the plasma rapidly inward, heating the ions and electrons. This shock heating phase is then followed by a slower adiabatic compression in which the plasma is compressed by increasing the magnetic field on a much slower time scale. In a more sophisticated concept, the "staged theta pinch", the shock heating and the adiabatic compression phases are driven by two circuits with different rise times and energy storage capabilities. During the adiabatic compression phase the plasma is heated to the ignition temperature of $\sim$5 keV where the fusion reaction begins. During the burning phase the plasma is further heated by the α-particles from the D-T-reaction (Eq. 1) and, since B is approximately constant during this time, the plasma expands against the magnetic field doing work which can be used for direct electrical conversion. Of course, the larger part of the energy released by the fusion reaction is carried off by the neutrons (Eq. 1).

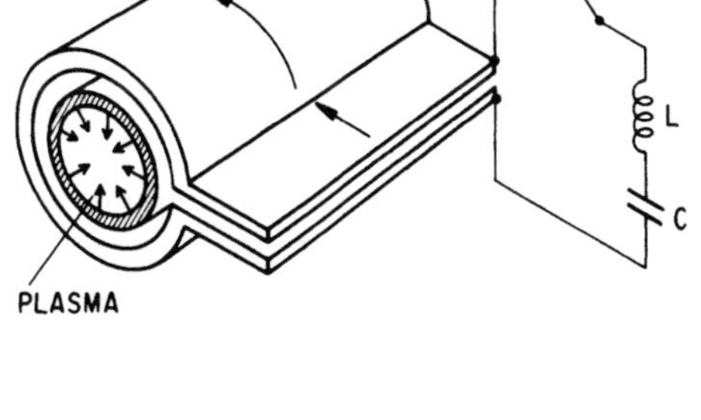

Fig. 3. Principle of operation of theta pinch

Both the open and closed (toroidal) configurations are feasible but the latter is preferred because of the elimination of plasma losses at the ends.

Theta-pinch concepts belong to the so called "high-β" devices since the ratio of the kinetic pressure of the plasma to the pressure of the external confining magnetic field is relatively high:

$$\beta = \text{const.} \frac{\rho T}{B^2} \quad (8)$$

where ρ and T are the plasma density and temperature respectively, B is the magnetic induction of the confining field[3, 4, 17]. The high-β devices have β values larger than $\gtrsim 0.1$, typically ~ 0.8. The significance of β in reactor performance is apparent from the dependence of fusion power density, P_{pl}, on the parameters of the plasma and of the confinement:

$$P_{pl} = \text{const.}\ \rho^2 <\sigma v> E_f = \text{const.} \frac{\beta^2 B^4 <\sigma v> E_f}{T^2} \quad (9)$$

where $<\sigma v>$ is the averaged product of the reaction cross section and particle velocity, E_f the fraction of the energy released by one fusion event which is deposited in the plasma[3, 4]. It is seen that the power density scales with ρ^2 (significant for inertial confinement) and, at a given plasma temperature, with β^2 and B^4 (significant for magnetic confinement). The Tokamaks, to be discussed later in this section, represent low-β devices with $\beta < 0.1$.

In the theta-pinch the plasma density is up to three orders of magnitude higher than in magnetic mirrors and Tokamaks and the confinement times are correspondingly shorter. Thus, the theta-pinch is inherently a pulsed device. The energy required for plasma heating and confinement is large as compared with the fusion energy release which leads to stringent requirements on energy transformation and storage systems[3]. The recent design study on the Reference Theta-Pinch Reactor suggest changes of the operating conditions which reduce the stored magnetic energy[18, 19]. Nevertheless, the magnetic energy storage still represents one of the most serious technological problems for the theta-pinch concept. Another difficult problem to be solved is the thermal loading and mechanical stress of the vacuum vessel wall.

The inertial confinement concepts utilize the idea of heating a pellet of D-T-fuel either by absorption of light from a powerful laser, a relativistic electron beam or a heavy ion beam to the ignition temperature in a time short compared to vaporization of the pellet. The reaction time must also be short compared with the confinement time to allow a sufficient burn up of the fuel for energy gain. In addition, the range of the 3.5 MeV α-particles has to be shorter than the pellet radius if their energy is to be efficiently deposited in the pellet.

In order to meet these conditions the core of the pellet must be compressed by a factor of 10^3 to 10^4 as compared to its solid or liquid density and heated up to ~ 10 keV on a time scale of ~ 1 nsec. The necessary power levels needed for the driver are of the order of hundreds of terawatts[2]. Availability of drivers providing such power at a sufficiently high repetition frequency, with reasonable efficiency and acceptable reliability represents only one of the key problems relevant to all the inertial confinement systems. Although significant progress has been achieved recently, the present experiments are far below the breakeven point[20, 21].

Several large experiments are planned which should demonstrate pellet gain in excess of unity, i.e. scientific feasibility. The experiments with laser drivers are under development e.g. at Lawrence Livermore Laboratory and at Los Alamos Scientific Laboratory (USA)[20]; those with relativistic electron beams at Sandia Laboratory, Albuquerque (USA)[21] and at Kurchatov Institute (USSR). They are scheduled to start operation in the early 1980's.

The Tokamaks represent today the most promising and best understood approach to fusion. The review paper by Bickerton[12] summarizes the status of Tokamak research in 1978. The physics of Tokamaks has been described by Artsimovich[22]. The name "Tokamak" is a Russian abbreviation of "Toroidal Kamera Magnetik", i.e. "Toroidal Chamber Magnetic".

The principle of operation and stabilization are shown in Fig. 4. The vacuum vessel contains the hydrogen isotopes at a pressure of the order of $\sim 10^{-3}$ mbar. Electric current I_t flowing through the toroidal windings, W_t, induces the toroidal magnetic field B_t. After the required value of B_t has been reached the gas is preionized, typically by an auxiliary high frequency generator, and subsequently the primary winding of the main transformer is connected to a powerful power supply. The primary electric current is schematically indicated as I_m in Fig. 4. Transformers with an iron or an air core are being used in present Tokamaks and, therefore, no details of the transformer design is included in the figure. The preionized plasma represents the secondary winding. The initial fast change of the primary current induces a time dependent magnetic flux, ϕ_m, which in turn induces a secondary toroidal current, I_{pl}, through the plasma. The vectorial sum of the poloidal magnetic field B_p of the plasma current with the toroidal magnetic field B_t form a helical structure of the magnetic confinement field in the Tokamak. The exact form of the helical structure is related to the $B_p : B_t$ ratio and, if properly chosen, provides the necessary stability of the plasma loop against magnetohydrodynamic (MHD) perturbations[22]. The so called "safety factor", q, given by Eq. 10:

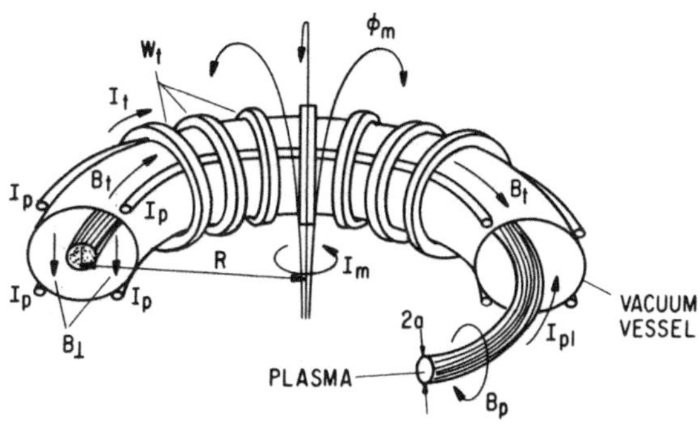

Fig. 4. Schematic view of Tokamak confinement and stabilization system

$$q = \frac{2\pi a/B_p}{2\pi R/B_t} = \frac{1}{A} \cdot \frac{B_t}{B_p} \tag{10}$$

determines which of the MHD modes will be stable. R and a are the major and minor radius respectively, Fig. 4. The aspect ratio $A = R/a$ in Eq. 10 is another important parameter characterizing the Tokamak. In present experiments and designs q is chosen to be >1 near the plasma center and $\gtrsim 3$ at the plasma edge.

It is inherent to the toroidal geometry that the confinement field decreases across the plasma cross section from the internal edge at $R - a$, towards the external edge at $R + a$. Thus, the plasma would be driven outward into the R-direction if there were not adequate stabilization. The latter is provided by a set of poloidal currents I_p properly adjusted to give an additional perpendicular magnetic field $B_\perp$ which stabilizes the plasma with respect to its radial movement. In operating Tokamaks the poloidal current, controlled by means of a computer programmed feedback system, determines the exact position as well as the shape of the plasma. It was shown theoretically that a vertically elongated plasma cross section should display better stability. An experimental device of this type is represented by Doublet III at General Atomic (USA).

As already mentioned, the induced current I_{pl} heats the plasma due to joule losses which are proportional to the plasma resistance. With increasing temperature the plasma resistance decreases and the ohmic heating becomes less effective. Theoretical analysis shows that the maximum temperature which can be reached in this way is below 2 keV, i.e. the plasma cannot enter the ignition region of ~5 keV[4]. Therefore,

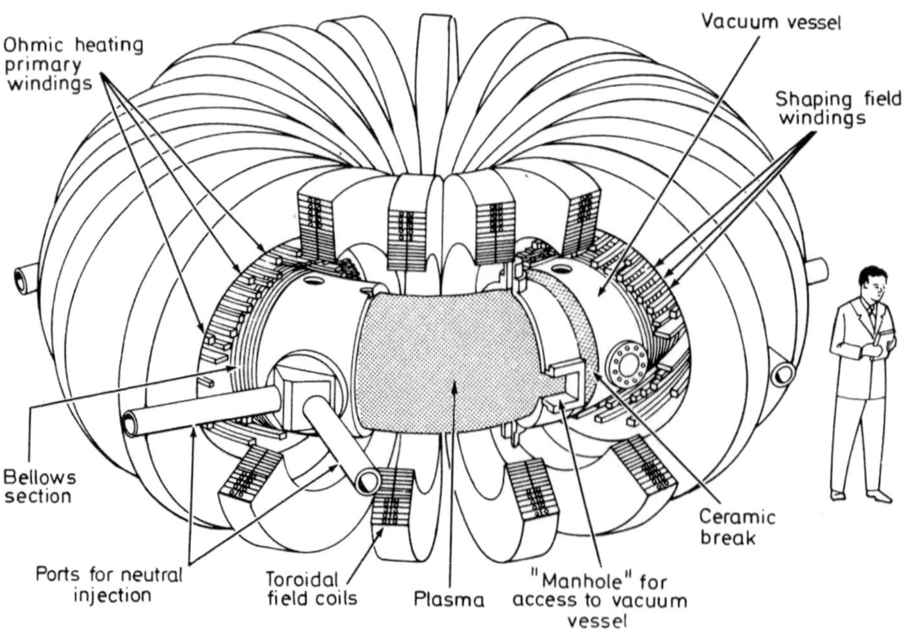

Fig. 5. Princeton Large Torus (PLT)[23]

auxiliary heating has to be applied immediately after the ohmic heating phase in order to raise the plasma temperature up to $\gtrsim 5$ keV. Neutral beam injection, adiabatic compression, and various rf auxiliary heating systems are presently being investigated. Once the ignition limit has been reached the energy of the α-particles from the fusion reaction (Eq. 1) is sufficient to compensate the total energy losses of the plasma in a stable regime[4)].

Operation of Tokamaks during the last 8–9 years has brought about a significant understanding of their physics and engineering, as well as an increase in their size, complexity and cost. Representative of the present Tokamaks is the Princeton Large Torus (PLT)[23)] shown in Fig. 5. Results obtained so far indicate that the next generation of large devices may enter the breakeven regime. The Joint European Torus (JET) now under construction at Culham (G. B.) gives an idea of the size of these machines (Fig. 6). The estimated costs are of the order of ~ 200,000,000 US dollars (1976). Future devices of this kind will be the TFTR (USA); JT-60 (Japan) and T-20 (USSR).

Fig. 6. Joint European Torus (JET) (With permission of Joint European Communities)

The large size of these machines and future Tokamak fusion reactors is dictated by several factors, of which, their relatively low power density is the most decisive.

In parallel with the experimental and theoretical studies a number of *conceptual fusion power reactor design studies* have been performed for each of the confinement systems. The objective of such work is to uncover and analyze in a self-consistent way the technological problems presented by a particular approach, to judge its feasibility and to point out the critical problems towards which future research work has to be directed. A brief summary of the past and present status of the Tokamak power reactor design studies is an appropriate way to get an idea of the wide range of the requirements for future technological development as well as for fundamental research related to this particular approach. Reactor design studies up to 1976 have been reviewed by Conn[24)] and a more recent summary can be found in Ref.[2)].

The conceptual design studies of the University of Wisconsin group has been chosen for discussion as an illustration of the past development and future trends in this field. Table 1 summarizes the significant parameters of three studies and Fig. 7

Table 1. Comparison of the major characteristics of three conceptual tokamak power reactor designs of the University of Wisconsin

Major characteristics	UWMAK I	UWMAK II	UWMAK III
Major radius R (m)	13	13	8.1
Plasma radius a (m)	5	5	2.7
Plasma height to width ratio	1	1	~2
Maximum toroidal field B_T (T)	8.66	8.3	8.75
Plasma current I_{pl} (MA)	21	15	15.8
Plasma power density P_p (MW/m^3)	0.78	0.78	2.1
Neutron wall loading P_n (MW/m^2)	1.25	1.16	2.50
Thermal Power P_{th} (MW)	5000	5000	5000
Electrical Power P_e (MW)	1700	1716	1985

References: UWMAK I[25)] UWMAK II[26)] UWMAK III[27)]

Fig. 7. Cross section of a noncircular Tokamak power reactor design-UWMAK III[27)]. Most of the details are self explaining or they were described in Fig. 4. The RF heating system heats the plasma from the maximum temperature attainable with ohmic heating (~1.8 keV) up to ignition. The "Separatrix" represents a boundary of the magnetic confinement vessel. Charged impurities leaving the first wall (carbon curtain and graphite ISSEC) do not cross the separatrix but they are moving along the magnetic lines into the poloidal divertor where they hit the particle bombardment plates and are either trapped or pumped away by the cryopumps. The blanket contains lithium for the tritium breeding. The inner and outer shields protect the superconducting magnets from radiation. Helium and lithium are used as coolants for the inner and outer blanket respectively

CROSS SECTION VIEW OF UWMAK III

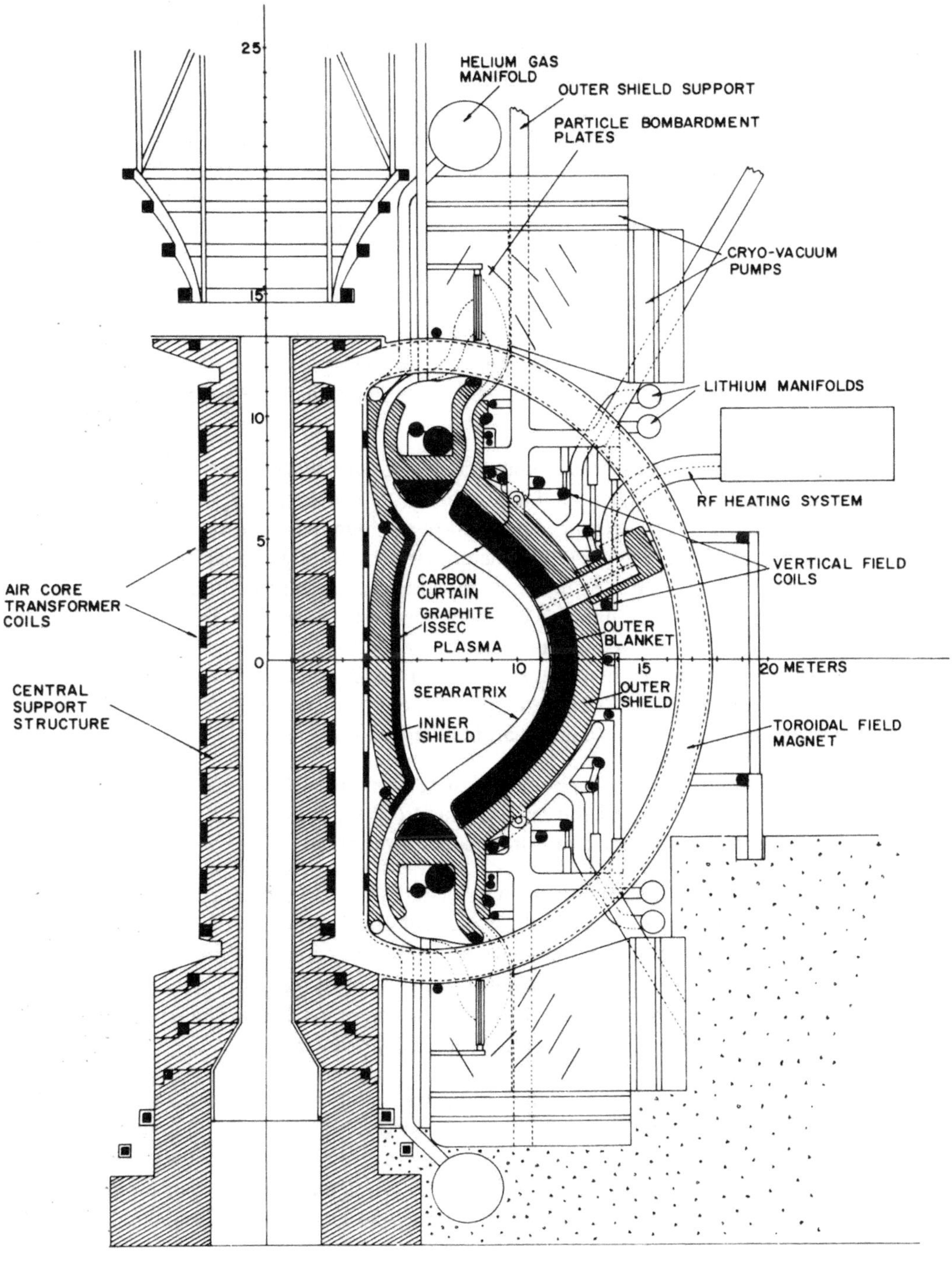

shows a cross sectional view of UWMAK III. The philosophy of these three designs differs mainly in the technology which is assumed to be available. UWMAK I[25] is conservative in terms of materials choices and operating conditions, whereas UWMAK II[26] uses a slightly advanced technology. Both designs are of the same size, plasma power density and comparable neutron wall loading. Structural material of the first wall is stainless steel 316, but UWMAK II uses a carbon curtain liner to reduce the level of high-Z plasma impurities as well as the neutron load on the metallic wall (see Sects. 3 and 6). Among other problems it turned out that the low plasma power density which results in a large reactor size for a given power has several serious drawbacks. For example, the requirements for some materials of construction which would be necessary in order to meet future energy demands by reactors of this type would be larger than the available resources and/or the world production[26].

These and further economic requirements have dictated the trend towards reducing the reactor size for a given thermal power. In order to achieve an adequate increase of the plasma power density higher values of β (Eq. 8) have to be used (Eq.9). A convenient way to achieve this requirement is to use plasmas of a noncircular cross section (Fig. 7).

The higher plasma power density and reduced size of UWMAK III[27] (see Table 1) result in a significantly higher loading of the wall by neutrons, charge exchange neutrals, and radiation. To sustain these conditions at high temperatures, a molybdenum based alloy, TZM, was chosen as the primary structural material. In addition, the UWMAK III design utilizes a number of materials and auxiliary equipment that assumes a rather advanced technology which has yet to be developed. In most respects the requirements cannot be unambigiously defined as yet. The recent paper on the selection of materials for first wall and divertor plates in magnetic and inertial confinement reactors represents one example of the present status[28]. With respect to Tokamaks, the problem of the materials for the first wall is twofold:

1) the wall has to sustain a high temperature, an intense flux of fast primary (Eq. 1), slow secondary (Eq. 6) and reflected neutrons for a sufficiently long time to enable economical operation of future fusion power plants;

2) the design of the first wall or the liner (such as the carbon curtain) together with the plasma confinement system, divertor, cold gas blanket, or other of the feasible components have to maintain a plasma sufficiently free of impurities to achieve and sustain ignition. This represents one of the most serious problems on the way to achieving the breakeven point in the near term machines. The remaining part of this chapter will deal chiefly with the impurity problem.

The interaction of the plasma with the first wall is, in the current Tokamaks, significantly reduced by the so called "limiter" which is typically a metallic ring mounted inside the torus in order to keep the plasma away from the first wall. In turn, the interaction of the plasma with the limiter is a significant source of plasma impurities. In many of today's Tokamaks the plasma impurity level is dominated by limiter erosion.

3 The Role of Impurities in Tokamaks

3.1 Effect of Impurities on Plasma Characteristics

As has already been pointed out, the plasma materials interaction represents one of many serious problems which have to be solved if controlled fusion based on Tokamaks is to become an adequate energy source.

Impurities are considered to be all elements present in the plasma except the D-T fuel. Their primary effect on the performance of the fusion plasma is two-fold:

1) dilution of the fuel resulting in a corresponding decrease of the plasma power density (Eq. 9);

2) enhanced energy losses which lead to the lowering of the plasma temperature.

Helium formed by reaction (1) is supposed to be removed due to recycling the fuel through an adequate recovery system.

The power radiated by a contaminated hot plasma due to impurities which are released from surfaces such as the first wall and limiter due to the plasma attack is greatly increased with increasing concentration and atomic number, Z, of the ions. The type of impurity enhanced radiation depends on the plasma temperature and on Z. At the operating temperature of ~ 10 keV in future Tokamaks, metals with Z $\gtrsim 20$ such as Fe, Ni, Cr, Zr, Nb, Mo, W, etc., are not fully stripped of electrons thus making recombination and line radiation the dominant energy loss mechanism under equilibrium conditions. These radiation losses appear to be significant even for lighter elements during the transient stripping process or at lower plasma temperatures when they are not fully ionized[23, 29–36].

To maintain a "burning" D-T plasma, the energy of the α-particles (Eq. 1) deposited in the plasma has to be larger than the plasma losses. This condition sets an upper limit to the maximum tolerable impurity concentration. A quantitative estimate assuming Bremsstrahlung to be the dominant process has been performed by Ekhartt and Venus[29], while a more detailed calculation has been done by Meade[35]. The results are summarized in Fig. 8. The solid line corresponds approximately to an impurity concentration at which the radiation losses due to Bremsstrahlung are equal to the power deposited in the plasma by the α-particles. To achieve ignition the impurity concentration must be less than the value given by the solid line. As already pointed out, Bremsstrahlung represents only a part of the total radiation loss. For heavy ions, the losses will be enhanced by line radiation which results in a significant lowering of the maximum tolerable impurity concentration as compared with the values given in Fig. 8[37]. Thus, from the point of view of plasma energy losses the low-Z impurities appear to be less critical than the heavy metals. Therefore low-Z materials such as carbon, boron- and silicon carbides, beryllium, boron, borides, and others are considered as coatings for first walls and limiters[25, 31, 38, 39].

The apparent advantage of using low-Z materials has several drawbacks however, such as their relatively high sputtering yields[40–42] and in case of carbon the chemical reactivity with atomic hydrogen[43–47]. A high low-Z impurity concentration, even if it is below the maximum tolerable limit, would result in a corresponding

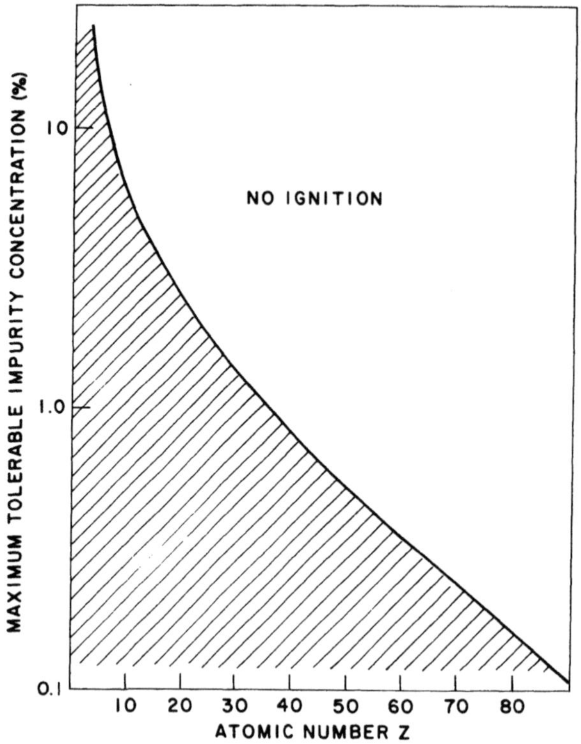

Fig. 8. Maximum tolerable impurity concentration of a D-T plasma (see text)

decrease of the fusion fuel and plasma power density (Eq. 9). Bohdansky, et al. have shown that the merit of low-or high-Z materials depends on the plasma edge temperature which is not yet known[48].

Plasma impurities may also have a beneficial effect by cooling the plasma edge in the vicinity of the first wall. But this is partially offset by the fact that the sputtering yields due to impurities are higher as compared with that of hydrogen. Consequently, a relatively small concentration of impurities can significantly increase the wall erosion and plasma contamination.

Of particular interest is the mechanism of impurity transport in the Tokamak plasma. Classical transport theory predicts an accumulation of heavy impurities in the plasma core with an adequate outward flux of hydrogen. However, various processes, such as plasma turbulence, can have an opposite effect and this problem is still a subject of theoretical investigation[49].

In some recent experiments in large Tokamaks with a rather clean plasma, obtained after extensive discharge cleaning, a hollow electron temperature profile has been observed which may be due to accumulation of heavy metal ions such as tungsten in the plasma core[50,51]. Similar T_e-profiles have been observed after introduction of tungsten atoms from an auxiliary source into the Tokamak plasma[52]. It is expected that experiments on large devices will help one to further understand the impurity transport mechanisms.

Finally the plasma-wall interactions result in erosion of the first wall which can, under long term operation, lead to significant changes of the wall thickness and composition. With primary particle fluxes of the order of $10^{19}-10^{20}$ ions m^{-2} s^{-1} and a sputtering yield of ~ 0.02, erosion rates of the first wall up to several mm per year are expected[36)]. Such values appear to be too high for reliable long term operation of a fusion reactor.

The fast neutrons will cause atomic displacement and transmutation reactions in the wall material. For example: after 20 years of operation, a niobium wall would contain 10 at % Zr, 0.06 at % Y, 0.28 at % He, and 0.5 at %H[53)]. The impact on the mechanical properties of construction materials due to displacement damage and radioactive transformation is under intensive study but does not properly fall under the subject matter of this chapter.

3.2 Fluxes to the Wall

Theoretical estimates of the erosion of the first wall and plasma contamination due to sputtering requires a knowledge of particle and photon fluxes to the wall, as well as data on the erosion yields. Sputtering will be discussed in a later part of this chapter. Here we shall briefly summarize some of the calculations done on primary fluxes in future fusion reactors. The calculations are rather uncertain because of the poor understanding of various parameters such as divertor efficiency, refueling, neutral beam heating, plasma temperature and density profiles, including the scrape-off layer in the case of divertor operated Tokamaks.

Most reliable are the data on neutron fluxes which are determined by the plasma power density, reactor geometry and structural parameters. Most of the conceptual power reactor design studies contain a detailed neutronics analysis giving the energy as well as the spatial distribution of neutron fluxes.

The fluxes of the primary 14.1 MeV neutrons are expected to be in the range of 10^{18} neutrons $m^{-2}s^{-1}$ whereas the fluxes of the reflected neutrons with energies below 14.1 MeV can be up to an order of magnitude higher.

The most essential plasma device characteristics that are needed in order to obtain impurity release rates are the fluxes of photons and of charged and neutral particles to the wall. It will be necessary to have detailed information on the energy spectra and fluxes to walls, limiters and beam dumps of thermal electrons and ions, photons, α-particles, runaway electrons, charge exchange neutrals, neutral beam and impurity neutrals and ions. The effects of sheath potentials, secondary electron emission and unipolar arcing need to be included in these calculations.

Fluxes must be specified during each phase of device operation including start-up, heating, burn, fueling, shutdown and failure modes. Plasma failure modes, such as instabilities which cause much of the stored plasma energy to be dumped in a short time may result in large particle or photon fluences to selected small areas of surfaces exposed to the plasma.

Although much has been done to improve the data base in these areas in the last decade, little is known about low energy ($E < 400$ eV) charge exchange neutrals, α-particle, and thermal and runaway electron fluxes.

Table 2a. Plasma parameters and power balance for a beam-driven D-T tokamak and an ignited tokamak reactor

Plasma parameters		TFTR[23]	UWMAK III[27]
Discharge duration	(s)	0.5	1900
Time between discharges	(s)	300	50
B (O)	(kG)	52	40
I	(MA)	2.5	15.6
R	(cm)	250	810
a	(cm)	85	400[a]
T_e (O)	(keV)	6	23
N_e (O)	(cm^{-3})	1×10^{14}	2.1×10^{14}
T_i (O)	(keV)	6	18
Z_{eff}		2	1
Dominant ions in plasma		T, D, (Fe, O)	T, D, (C)
ions in beam		D	
Power balance			
Fusion events	(MW)	25	5000
Ohmic heating	(MW)	0.18	0.9
Auxiliary heating	(MW)	32 (0.1 s duration)	200 (1.4 s duration)
(Neutral beams – TFTR; RF-UWMAK)			
Loss to wall surface	(MW)	47	440
Charge exchange (E < 300 eV)	(W/cm^2)	0.05	0.01
Charge exchange (E > 300 eV)	(W/cm^2)	3	0.5
Radiation	(W/cm^2)	8	4
Conduction	(W/cm^2)	15	0.01
Alpha particles (3.5 MeV)	(W/cm^2)	1	0.1[a]
Neutrons (14 MeV)	(W/cm^2)	20	250
Loss to limiter	($MW : W/cm^2$)	10:2000[a]	0
or divertor	($MW : W/cm^2$)	9	725:600

[a] Estimated

Table 2b. Fluxes of energetic particles and photons to the first wall[6]

Particles	Energy	Flux
D^0, D^+, T^0, T^+	$1 - 5 \times 10^5$ eV	$10^{18} - 10^{20} m^{-2} s^{-1}$
He^+	$1 - 3.5 \times 10^6$ eV	$10^{16} - 10^{18} m^{-2} s^{-1}$
Impurities	$1 - 10^4$ eV	$10^{15} - 10^{17} m^{-2} s^{-1}$
e^-	$1 - 10^5$ eV	$10^{17} - 10^{19} m^{-2} s^{-1}$
$h\nu$	$10^{-3} - 10^4$ eV	$< 100\ W\ cm^{-2}$
D^0, T^0	<1 eV	$\lesssim 10^{22}$[a] $m^{-2} s^{-1}$

[a] see text

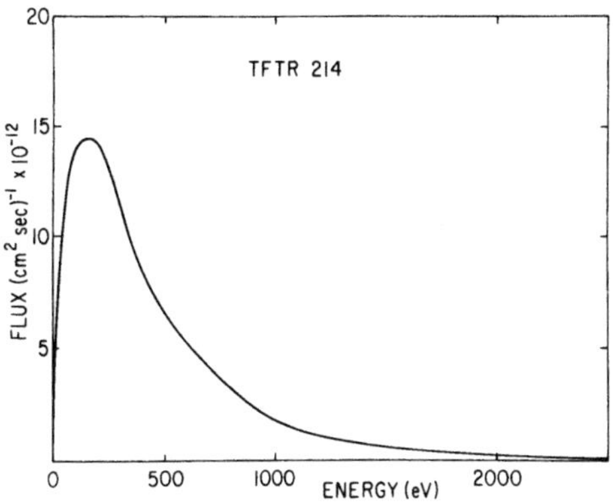

Fig. 9. Energy spectrum of the calculated flux of charge exchange neutrals which escape from TFTR immediately after neutral beam heating[23)]

The plasma parameters and power balance for a beam driven D-T Tokamak and an ignited Tokamak reactor are listed in Table 2a for illustrative purposes. The ranges of photon and particles fluxes to be expected on the first wall are listed in Table 2b.

It is of interest to note that only recently has it been recognized that in large, dense Tokamaks, the energies of the ions in the plasma core are much higher than the energies of the charge exchange neutrals that reach the wall. The latter are more nearly representative of plasma edge temperatures as illustrated in Fig. 9 which shows the energy spectrum of the calculated flux of charge exchange neutrals that escape from TFTR immediately after neutral beam heating. The maximum in the distribution occurs at 200 eV, much lower than the 10 keV ion temperatures in the plasma core. This important finding has dramatically reoriented much of the surface physics work from sputtering yield measurements done at 10 keV energies to yield measurements in the sub-keV range.

Selfsputtering, i. e. the sputtering of the first wall by the impurity ions can have a dramatic effect on these data because of the large sputtering yields of heavy ions as compared to that of hydrogen isotopes and helium. The resulting enhanced plasma contamination can lead to changes of many parameters (e.g., temperature and density profiles) and it can also influence the local as well as the overall plasma stability. Under certain conditions a positive feedback could develop leading to catastrophic instability and interruption of the plasma pulse.

Only very limited data are available on the density and temperature of the gas in the shadow of the limiter. The actual composition of this gas layer depends on many parameters of the Tokamak operation as mentioned above, but the degree of ionization is generally rather low there. Since the gas density is in the range of $\sim 10^{19}\,m^{-3}$, the flux of particles with energies below 1 eV can reach values of $\sim 10^{22}\,m^{-2}s^{-1}$. Such an estimate should be modified in devices using gettering surfaces but no exact values are available as yet.

3.3 Impurity Release Mechanisms

In concluding this section we should like to give a brief overview of the processes involved in the release of impurities due to plasma-wall interactions. These processes will be discussed in some detail in Sect. 5.

The initial contamination of a Tokamak plasma used to be dominated by oxygen and carbon which are released from the wall and from the limiter due to desorption (see Sect. 5.5.). These impurities were of primary concern in most machines until the discharge cleaning technique came to be routinely used[54]. In addition to the various desorption processes, a number of other impurity release mechanisms need to be considered. Among these are:

- physical and physichemical sputtering
- chemical erosion
- vaporization
- blistering and flaking due to ion bombardment
- arcing
- emission of chunks (metallic dust particles) into the plasma
- embrittlement (fracture by thermal shocks, etc.).

The relative importance of each of these processes depends on the actual conditions in a particular Tokamak, on the materials used, the plasma parameters, and the time evolution of impurities. The plasma-wall interaction can be rather nonuniform as is, for example, observed on the limiter[55,56]. In present Tokamaks, plasma contamination seems to be, after intense discharge cleaning, dominated by erosion of the limiter, the most important mechanism being evaporation, arcing, cracking and sputtering. Sputtering appears to dominate in the quiet phase of the discharge. The relative contribution of arcing to plasma contamination depends on the surface conditioning and it occurs only if the limiter is biased negative with respect to the surrounding "scrape-off" plasma. Evaporation takes place due to the local surface heating by fast electrons[57]. Sputtering of the first wall by ions and fast charge exchange neutral atoms will become an important impurity release mechanism in large devices and reactors[41,51,58].

4 Hydrogen Isotope Recycling

Hydrogen isotope recycling is largely responsible for impurities introduced by a variety of sputtering and erosion processes occurring at the first wall and limiter[59]. Solutions to the "impurity problem" therefore involve schemes for modifying the recycling processes and minimizing the erosion rates at the surfaces facing the plasma. All of these phenomena will assume even greater importance in future confinement devices and reactors which require hotter plasmas, longer confinement times, greater heating power and higher purity.

Surface analytic and spectroscopic studies have shown[60] that trapping and subsequent re-emission of hydrogen isotopes from walls dominate the fuel balance in ungettered discharges. Hydrogen isotope recycling strongly affects plasma profiles, especially at the edge, hence substantially modifying impurity influxes. In D-T-burning devices such as TFTR, wall recycling will strongly influence tritium inventory, which must be held to within well defined limits[51,61]. Therefore, to the many other factors which are important for the selection of a first wall material, must be added that of tritium retention.

4.1 Reflection

The first wall in a magnetic fusion reactor is bombarded by positive ions, electrons, neutral atoms, and photons from the plasma which initiate a myriad of complex interactions contributing to the recycling problem. Ions and neutral atoms may be reflected or back-scattered from the wall with a change of charge, atomic state, and/or energy. The relative magnitudes of these effects which actually occur vary depending on the ion species, energy, and angle of incidence, the target species and condition, and any near-surface alterations of the material resulting from plasma bombardment, ion implantation, thermal cycling, etc. Reflection coefficients for hydrogen ions and neutrals and their angle, energy, excited state, and charge state distributions are needed for plasma modeling calculations. As an indication of the importance of the problems, the particle confinement time in present devices is only 1/5 to 1/10 of the discharge time; this means that on the average, all plasma particles hit the first wall and are recycled into the plasma 5 to 10 times during one discharge. In future devices the importance of these problems will depend somewhat on the exact reactor configuration parameters, but a broad understanding of the recycling phenomena must be established under the actual operating conditions of each reactor.

The following discussion is taken from a review article by R. Behrisch[62)].

Total backscattering yields and backscattered energy as well as energy and angular distributions of the backscattered atoms have been calculated analytically[63, 64)] and by computer simulation programs which are able to follow the individual ion trajectories[65–67)]. Experimental determination of backscattering yields and distributions are difficult, as in the energy range of interest most of the backscattered particles are neutral. The most successful way of ionizing the neutrals is a gas filled stripping cell, as also used in neutral particle measurements in plasma experiments[68)]. Such a cell could be calibrated to energies as low as 130 eV[69)].

The results for total backscattering yields and backscattered energy of hydrogen ions on different materials are shown in Fig. 10[63–65, 70–72)]. The values are plotted as a function of a dimensionless reduced energy ϵ given by[73)]:

$$\epsilon = \frac{M_2}{M_1 + M_2} \frac{a}{Z_1 Z_2 e_o^2} E \tag{11}$$

where M_1, Z_1 and M_2, Z_2 are the masses and charge numbers of the incident ions and the target atoms, respectively, e_o is the elementary charge ($e_o^2 = 14.39 \times 10^{-10}$ eV · m; 'a 'is the Thomas Fermi screening length given by: $a = 0.486 \times 10^{-10} \times$ $\times (Z_1^{2/3} + Z^{2/3})^{-1/2}$ and E is the incident energy in eV. In the two lower scales of Fig. 10 absolute energies for stainless steel (Fe) and Mo are introduced. The measured numbers show reasonable agreement with the calculated values only for some materials.

Figure 11 shows an energy distribution of the backscattered positive, negative and neutral atoms[70)]. It can be seen that more than 90% of the backscattered atoms are neutral in this low energy range, while negative and positive ions have nearly equal intensities. All spectra indicate that the backscattered particles may have an energy up to nearly the incident energy, while a maximum is seen at around 1 to 2 keV. The position of this maximum does not depend strongly on the incident energy. This

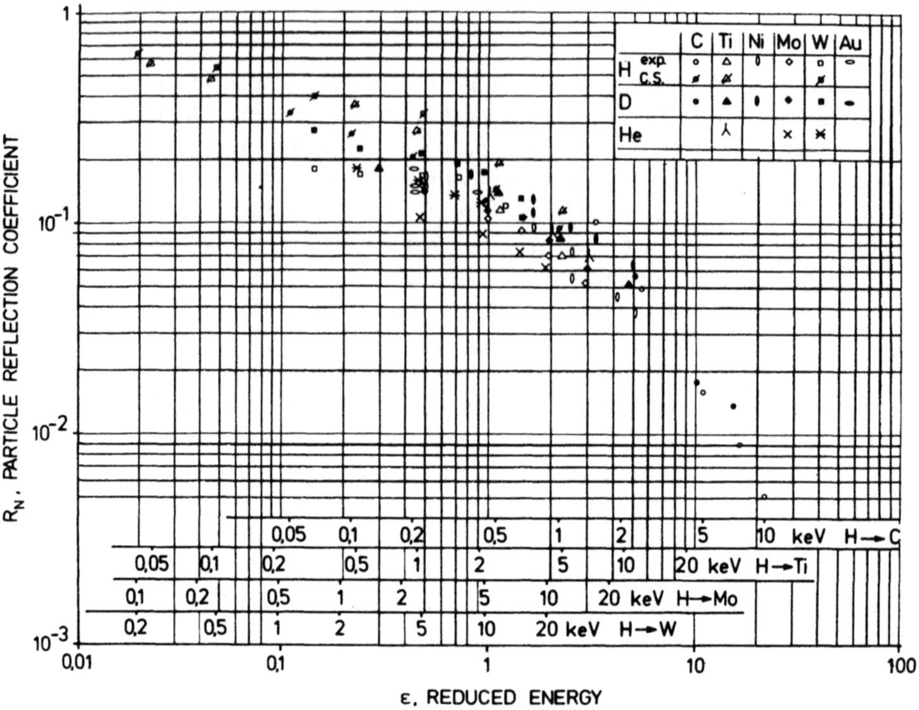

Fig. 10. Particle and energy reflection coefficients as determined by different authors

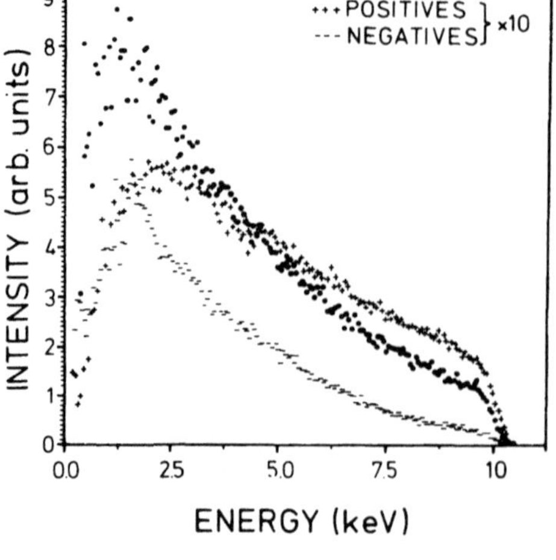

Fig. 11. Energy distribution of the positive, negative and neutral atoms backscattered from a stainless steel (304) surface for 10 keV D bombardment. The ion beam was at normal incidence and the backscattered particles had been observed at an angle of 45 deg. to the normal[70)]

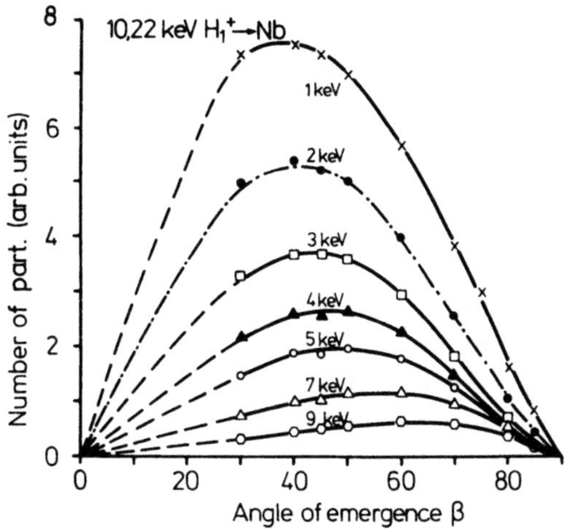

Fig. 12. Angular distribution of protons backscattered from polycrystalline Nb target into different energy ranges[74]

means that for lower bombarding energies the maximum is shifted closer to the incident energy[65, 66, 70].

An example of an angular distribution of the atoms backscattered into different energy ranges for normal incidence is shown in Fig. 12[74]. While those atoms backscattered with low energy display nearly a cosine distribution, the high energy particles show predominant backscattering at glancing angles (80° to the surface normal).

The implication of these results to a fusion reactor is that part of the ions leaving the plasma are directly backscattered from the first wall into the plasma as neutrals with energies up to near the incident energy. They may penetrate deep into the plasma and create more energetic charge exchange particles from the plasma than those neutrals released with thermal energy from the first wall.

4.2 Gas Re-emission and Trapping

When an energetic ion is incident on a surface it has a probability of penetrating into it and slowing down to thermal energies, or of being backscattered out of the surface with an appreciable fraction of its initial energy, as already discussed. Those ions not directly backscattered become neutralized and come to rest in the solid, thus contributing, for example, to the T-inventory. After slowing down they generally occupy interstitial positions and may diffuse further. Depending on the solubility, the diffusibility and the barrier at the surface, they may either diffuse into the bulk of the solid, they may be trapped in the implanted layer (generally at damage sites)[75], or they may in part diffuse to the surface and be re-emitted into the plasma as cold atoms. Re-emission and trapping processes must be quantified in first wall materials to establish how surface properties affect the phenomena.

During repetitive machine operation a considerable quantity of gas can be built up in the surface layers of the first wall, and re-emission rates should readily approach 100% within a finite number of discharges. The exception to this is for metals which are strong hydride formers and exhibit high solubility for hydrogen isotopes (e.g.,

Ti, Zr), where re-emission should be very small, at least near ambient temperatures. For example, titanium hydrides can be formed in a low pressure hydrogen discharge at ~ 700 °C. Hydrogen and helium interactive effects can be large. Strong enhancements in the trapping of hydrogen isotopes have been observed by a number of investigators after helium or other ion pre-damage[76, 77]. It was observed that the hydrogen tends to decorate the damage depth profiles of the damaging ion.

The release of the trapped hydrogen isotopes is temperature dependent, so that temperature cycling in the presence of these interacting effects could be significant. In addition, the retention and mobility of hydrogen is dependent on the near-surface condition of the material. The presence of cold work was shown to increase the hydrogen isotope trapping in the absence of other ion damage[78]. This shows that the influence of the metallurgical state of the material is important to the trapping effect. Measurements in insulator and semiconducting materials have shown that hydrogen is immobile in highly damaged or amorphous materials but extremely mobile under positive ion irradiation in polycrystalline materials[79].

The remaining part of this section is largely based on a discussion of the subject by W. Bauer[80].

Hydrogen isotope retention in stainless steel is extremely important from both the plasma physics and environmental points of view. Recent measurements[60] on Tokamaks have demonstrated that trapped hydrogen isotopes are desorbed from the first wall during a discharge and dominate the plasma refueling.

Data from existing devices on the amounts of H isotopes trapped in the wall are now becoming available. For example, in Fig. 13 are data from PLT exposures for stainless steel and Si samples[81]. One notes that nearly all the D trapped in stainless

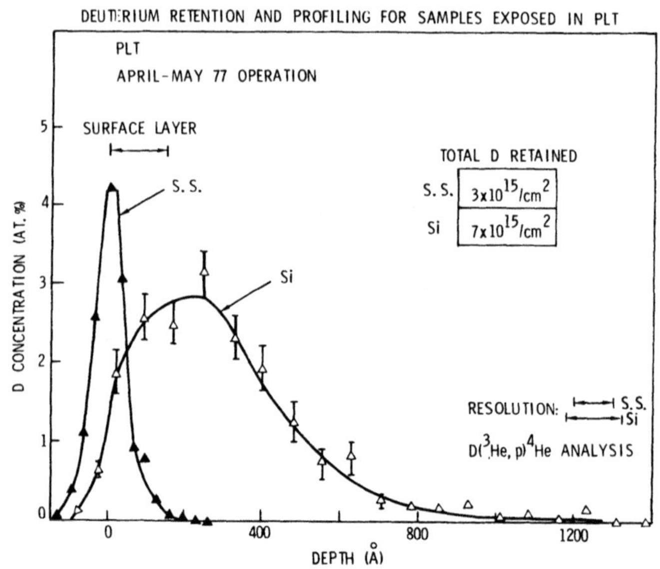

Fig. 13. Nuclear reaction profiling with D (^{3}He, p)^{4}He of deuterium in silicon and stainless steel exposed to two months of PLT discharges[81]

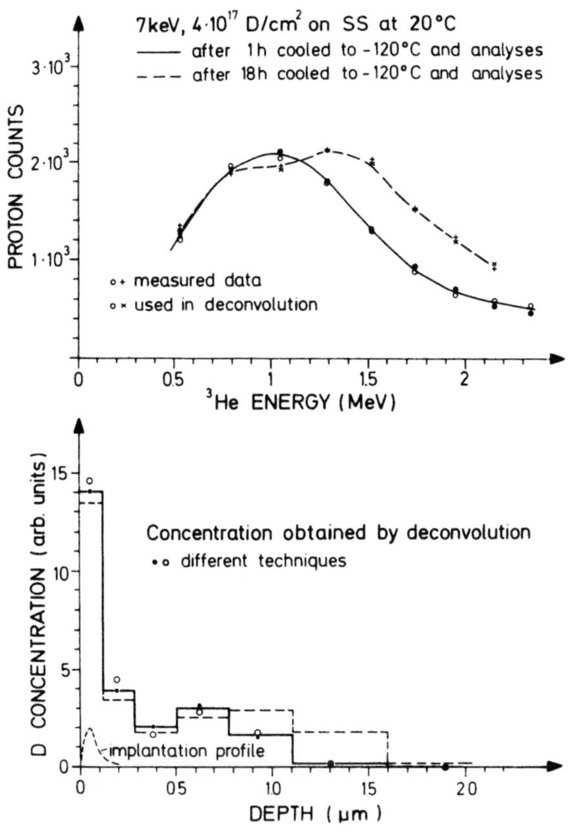

Fig. 14. Nuclear reaction profiling with D (^{3}He, p) ^{4}He of deuterium in stainless steel implanted with 14 keV D_2^+, and quenched to -120 °C after one hour retention at room temperature[82)]

steel is in the surface layer which is predominately carbon. In Si, bulk trapping may also have occurred.

Several recent experiments have been conducted on the hydrogen retention characteristics of stainless steels. Figure 14 shows the depth profile of deuterium in 316 stainless steel, measured by nuclear reaction profiling one hour after room temperature implantation with 14keV D_2^+ to a fluence 4.10^{21} D m^{-2} [82)]. Significant amounts of deuterium are observed at depths well beyond the incident particle range, indicating that hydrogen diffusion has occurred. Deuterium retention in the near surface region is also visible.

In Fig. 15 hydrogen desorption data from linear ramp thermal anneals of oxidized stainless steel are presented.

The samples were exposed to milliampere currents of 300 eV hydrogen for several hours. The desorption data, taken 30 min and 17 h after bombardment show that a large amount of hydrogen (up to 6.10^{20} H m^{-2}) is retained in or near the sample surface, and that the hydrogen is easily desorbed. Similar experiments from other laboratories give general agreement with these results[83)].

Experimental results indicate that hydrogen isotopes reside in the near surface region as well as in the bulk stainless steel after low energy bombardment. The total retention of hydrogen isotopes is relatively low, indicating that tritium retention in

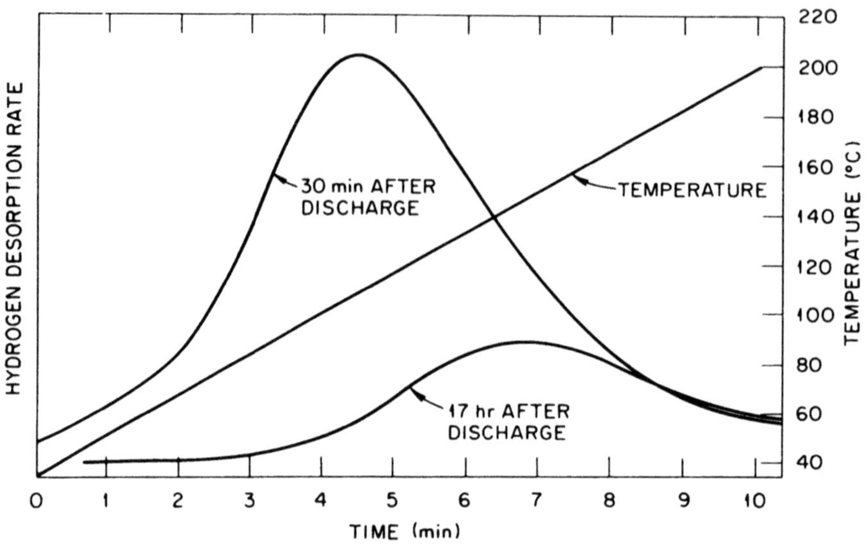

Fig. 15. Linear ramp thermal desorption of oxidized stainless steel samples implanted with 300 eV hydrogen at room temperature (Clausing, R. E., et. al.: in: Ref.[39], p. 573). The thermal desorption technique shows that: (1) a large amount of hydrogen is adsorbed in or near the sample surface, (2) the hydrogen is easily desorbed

stainless steel first walls of early D-T devices should be within acceptable environmental limitations. However, the hydrogen isotopes trapped in the oxide or radiation damaged layer at the surface is a major source of the hydrogen isotopes recycled from the wall during a Tokamak discharge. In Tokamak devices operated with the first wall at room temperature, release of trapped hydrogen isotopes from uncoated stainless steel surfaces during a discharge appears to be inevitable.

The H isotope trapping behavior in other materials is different. The results of Langley et al.[84] and Erents et al.[85] present a fairly complete and consistent picture of trapping and desorption effects during the implantation of graphite with low energy (< 20 keV) deuterium ions. By depth profiling the implanted deuterium it was found that the local deuterium concentration reached 0.6 atom fraction at a fluence of 10^{22} D m^{-2}. Above this fluence the deuterium concentration remains constant. The trapping efficiency was found to be 100% for fluences up to 10^{22} D m^{-2} but decreased with increasing fluence.

Recent improvements in PLT[86] involving titanium gettering have revived interest in active metal coatings on first walls. The titanium gettering between discharges in PLT improved the base pressure of the Tokamak, and may have influenced the recycling of hydrogen between the plasma and the first wall. Figure 16 shows the results of gas re-emission measurements[87] and their relation to the trapping coefficients of several active metals as a function of temperature. The data show almost 100% trapping at room temperature for all three cases. However, before the concept of gettering can be applied to D-T devices such as TFTR, a technique for removing the trapped tritium from the Tokamak must be devised. As seen in Fig. 16, titanium must be heated to $> 500°$ K before the hydrogen isotopes are rapidly released.

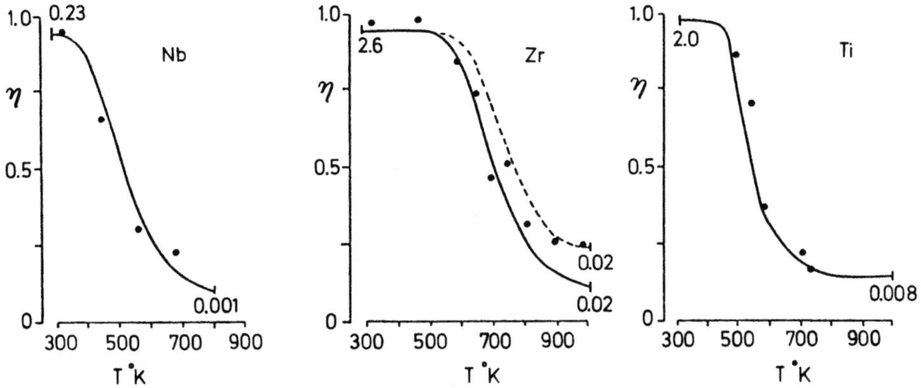

Fig. 16. Trapping coefficient for Nb, Zr, Ti targets bombarded with a 50 μA (0.7 mA/cm^2) beam of 18 keV D^+ for 1000 s as a function of temperature. The numerals alongside the curves give the atomic concentration of the trapped ions at the centre of the focal spot. The solid curve for Zr is for a value of Q_2 = 8250 cal/mole and the broken line for Q_2 = 6940 cal/mole. The experimental points are for a nominal current of 50 μA[87])

The effects discussed above are important, not only for the plasma first wall, but even more so for neutral beam injector components and beam dump areas, as well as divertor surfaces. (The role of divertors in the control of impurities is discussed in Sect. 6.1.) For divertor operation, it is desired that the collector surfaces be 100% gas absorbers. Surface physical and chemical conditioning of divertor surfaces could provide longer life and greater gettering surface areas. Re-emission from divertors and subsequent backstreaming to the plasma need to be taken into account. If this were to lead to 100% re-emission, it would correspond to a particle re-emission rate of 10^{22} $m^{-2}s^{-1}$. If only 1% of the ions backstreamed into the plasma it could result in a neutral charge exchange flux to the wall as high as 10^{20} m^{-2} s^{-1}, assuming 100% charge exchange probability. Because of the importance of trapping both for gettering and divertor operation, the characteristics of trapping surfaces are discussed in more detail in Sect. 6.4.

Re-emission rate data are needed during bombardment, as a function of time. Fluxes should be of the order 10^{20} particles $m^{-2}s^{-1}$, and target temperatures should range up to 800 °C. Targets should include samples that have undergone radiation damage by hydrogen and helium ions and by neutral atoms.

For the area of gas trapping, information is required on the conditions for hydride formation, the rate of such formation, and the stability of the resultant compounds under the extremes of operating environments. In the case of chemical trapping, data on hydrogen retention should be obtained on damaged samples as a function of temperature and particle energy. Considerable data exist on the stability of bulk hydrides and for the solubility of hydrogen in metals, but essentially no data exist on the nature and stability of compounds formed as result of plasma-materials interactions. Although considerable information is available on gas release from ion bombarded samples, there is very little information on trapping and release mechanisms.

5 Impurity Release Mechanisms

Plasma contamination by impurities coming from walls, limiters, divertors, beam dumps, or background gas has become a problem of increasing importance as confinement times, particularly in Tokamaks, have increased. The possibility that impurities can concentrate in certain regions of Tokamak plasmas has placed an additional premium on understanding the phenomena leading to plasma contamination and ultimately devising means for drastically lowering impurity levels. The deleterious effects, particularly of high-Z impurities, on the performance of magnetically confined fusion machines stems from their effectiveness in strongly enhancing power losses from thermonuclear plasmas due to increased line, recombination, and bremsstrahlung radiation. (See Sect. 3.) The important phenomena giving rise to plasma contaminant release are discussed in detail here, although they were surveyed in a general way in Sect. 3.3.

5.1 Sputtering

5.1.1 Physical Sputtering

Accurate measurements of sputtering yields from particles escaping the plasma and impinging on first wall materials are required in order to assess the importance of physical sputtering to the impurity generation problem, to the long-term integrity of first-wall materials and to provide data for plasma modeling calculations. The areas of sputtering where information is most needed for the fusion program (low-energy light ions on alloys, compounds, or ion-implanted surfaces) are among those least understood. There are relatively few measurements of sputtering yields for the kinds of particles which are expected to impinge on the first wall, and those measurements which do exist are in poor agreement with existing theories[42)].

The low sputtering yields ($<10^{-3}$) dictate severe experimental requirements for improved results such as ultrahigh vacuum and meticulous surface characterization. Nevertheless, these data must be obtained and reconciled with theoretical models in time to provide modeling input for future high-power reactors, where sputtering may become the dominant source of impurity introduction. Accurate sputtering yields for known heavy-ion plasma contaminants are also essential. Even small concentrations of heavy-ion impurities can initiate severe problems because of their large sputtering yields.

Since any particle sputtered from the first wall will extract energy from the plasma in the course of being ionized, excited, dissociated, or when it undergoes radiative recombination, it is vitally important to have accurate measurements of the charge, energy, quantum state, molecular state, and angular distributions of sputtered particles. Very little information exists on the energy and angular variation of sputtered species, and virtually none taken in-situ during Tokamak operation. Information concerning interactive effects on sputtering, including sputtering of alloys and compounds, preferential sputtering, and the effects of near-surface radiation damage and diffusion on sputtering, is required. A more complete understanding is needed of the effects of

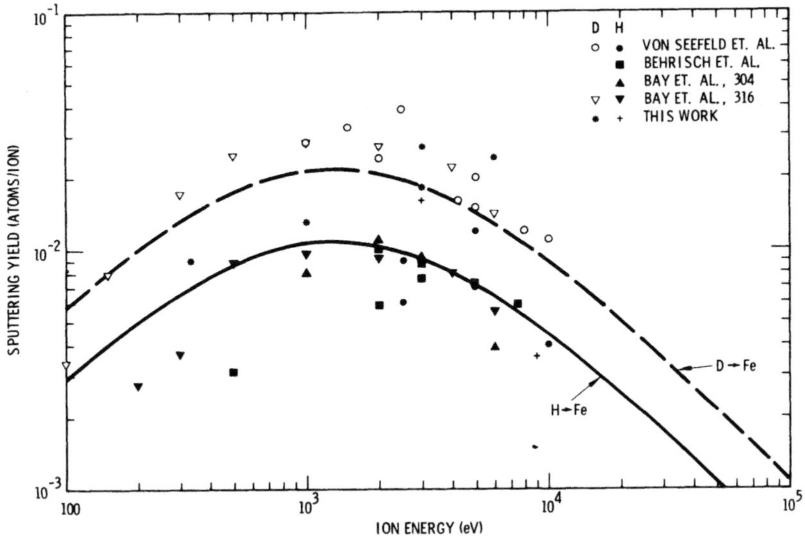

Fig. 17. Sputtering data for hydrogen and deuterium on 304 and 316 stainless steel[88]. In addition to the new data of this work, the data of Von Seefeld (○,●) [Von Seefeld, H., et. al.: J. Nucl. Materials *63*, 215 (1976)]; Behrisch et. al. (■) [Behrisch, R., et. al.: J. Nucl. Materials *60*, 321 (1976)]; and Bay et al. (▲, △, ▼) (Bohdansky, J., Bay, H. L. and Roth, J.: in: Proc. 7th Int. Vacuum Congr. and 3rd Int. Conf. Solid Surfaces, Vienna, 1977, R. Dobrozemsky et. al. (Eds.), F. Berger and Sohne, Vienna, 1977, p. 1509) have also been included

surface structure and surface chemistry on the sputtering process. Finally, theoretical models suitable for use in predictive modeling codes need to be developed and verified, particularly as regards the theory of light ion sputtering.

Much of the available data for low energy hydrogen isotope sputtering is for higher (>26) Z materials at energies of 100 eV and above. Collected data are shown in Fig. 17 for H and D sputtering of 304 and 316 stainless steel[88]. The data form a rather consistent envelope and the semi-empirical curve developed by Smith[89] describes the data rather well, except for the lowest energies.

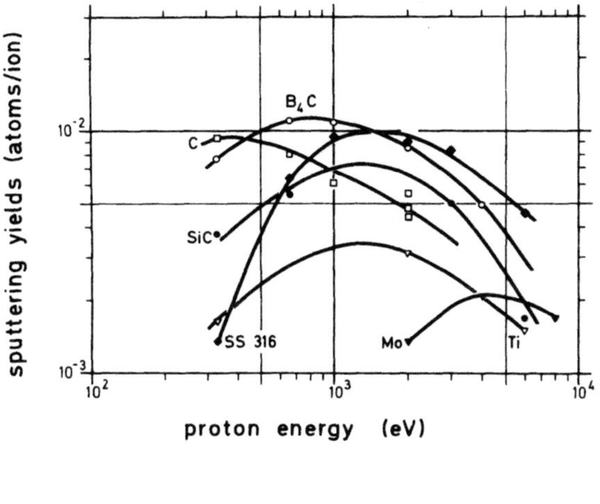

Fig. 18a. Sputtering yields for different materials in dependence on proton energy. All values are taken at normal incidence and target temperatures between 50 °C and 150 °C[42]

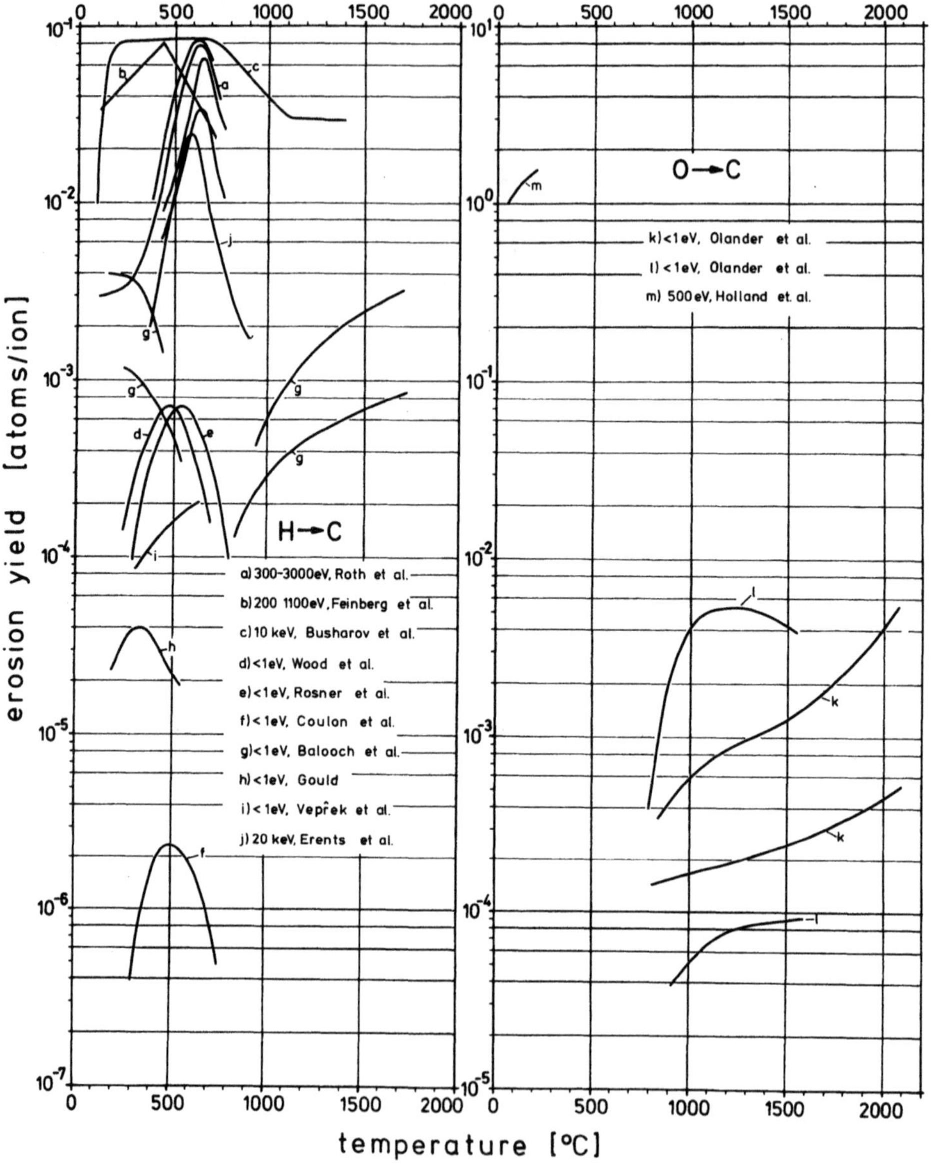

Fig. 18b. Temperature dependence of erosion yields for protons and oxygen ions on carbon. The energies of impinging ions as used by various athors are indicated in the insert (from Ref.[41])

For lower Z materials, less data are available than for stainless steel. In Fig. 18a are shown the data of Bohdansky et al.[42], for a variety of low-Z materials plus 316 stainless steel and molybdenum. From these data, a trend in the energy dependence of the sputtering yields can be discerned. The position of the maximum of the yield as a function of energy shifts towards lower energies as the atomic number of the sputtered species decreases. The peak in the carbon sputtering yield is reached at an energy on the order of 100–300 eV. It appears that due to the higher sputtering yield at low energies a low Z wall may not result in lower impurity radiation for very low edge temperatures[48)].

Most of the sputtering data are for pure materials. In fact, most investigators have attempted to avoid surface contamination in order to make measurements truly representative of the substrate material. In a Tokamak reactor the wall will be sputter cleaned and hence these data are relevant. The actual conditions in today's operating Tokamaks are more complex. For example, carbon and other materials layers have been shown to build up on the surface of walls, and hence it is desirable to have sputtering data on samples with impurity layers[195)].

5.1.2 Physichemical Sputtering

The typical wall material probably does not present a rigorously clean surface to the plasma but has at least partial oxygen and carbon coverages. The sputtering rate and nature of the ejected species will be a strong function of surface composition. It is important to know the sputtering coefficient for removal of O and C absorbed on the machine walls. This has a serious bearing on the effectiveness of discharge cleaning. Included here is the question of the molecular state formed by the sputtered projectile and a simultaneously sputtered atom; for example MoO is known to be formed in significant amounts when O^+ sputters pure Mo[90)]. To encompass these various chemical effects on the basic mechanism of physical sputtering the term "physichemical sputtering" has been used.

To fully understand the influence of sputtering on plasma behaviour one would need to study sputtering of realistic surfaces carrying adsorbed, chemisorbed, and implanted impurities. One must monitor not only the gross material removal rate but also the nature of the molecular and atomic species which are ejected. It is particularly important to include sputtering by plasma impurity species; while their flux may be lower than that of hydrogen, their sputtering coefficients are much higher, so that their contribution is significant.

A practical problem is that the sputtered chemical complexes often exhibit only a transitory existence. Traditional mass spectrometric techniques provide information on gross removal but little understanding of the mechanism involved. Data requirements in this area include gross removal rates, nature of ejected species, and changes to surface stoichiometry. In Sect. 6.5.1., the influence of surface chemistry on the nature of the sputtered species, and in particular on secondary ion fractions will be discussed. The role of surface chemistry and surface modification procedures on impurity control will be elucidated there.

5.2 Chemical Erosion

An operating CTR (Controlled Thermonuclear Reaction) device produces an extremely reactive chemical environment, the most important aspect of which is the highly reducing atmosphere resulting from the active forms of hydrogen that are produced. These can react readily both with adsorbed species and with the material of the wall itself. The former is expected to be more limited in magnitude, since it is restricted to monolayer quantities. The problem of reaction with adsorbed species would decrease as discharge cleaning techniques are more fully developed and as wall temperatures increase during machine operation. Chemical reactions with wall material will tend to maintain an oxide surface on stainless steel and insulators (e.g. SiO_2, Al_2O_3 and others) in the reduced state and thus modify processes such as sputtering and gas re-emission.

The question of surface chemical reactivity is critically dependent on the chemical nature and composition of the wall and the chemical nature and energies of the particles arriving at the wall. It will be necessary to develop experimental techniques to monitor the changing chemical composition of the first wall resulting from the flux of catalytic reactive plasma particles. Since the presence of active forms of hydrogen is inherent in all fusion devices, the potential for problems associated with surface chemical reactions of these species will always be present, only the form and magnitude of the problem will change.

Chemical sputtering occurs when a bombarding species forms a volatile compound with the target material. An obvious example is the bombardment of carbon with hydrogen where at elevated temperatures formation of CH_4 occurs with a simultaneous large rise in sputtering yield[91, 92]. Figure 18b shows the dependence of erosion rates on temperature as measured by various authors[441]. There is ample evidence that carbon reacts readily also with thermal hydrogen atoms, the steady state concentration of carbon in the gas phase reaching up to 5–7 at % even if the bombardment by ions is negligible[96]. Generally, the sputtering yields are a factor of about 10 to 100 higher than the reaction probability for thermal atoms. This difference is apparently due to the difference between the trapping probability for energetic ions and the sticking probability for thermal atoms[134]. During sputtering, radiation damage of the target is produced within a layer of thickness comparable to the projected range which can influence the reactivity of the material. For example, annealed pyrolytic graphite shows a relatively low reactivity towards thermal hydrogen atoms which is, however, increased by more than an order of magnitude if radiation damage has been produced by irradiation with ions in the MeV range[206]. More recent measurement which were performed by the Zürich group have shown that different carbon materials including various kinds of pyrolytic graphite, isotropic nuclear graphite and glassy carbon which were exposed to the same fluence of ions (2 MeV – He^+) display the same reactivity with atomic hydrogen within about a factor of 2, regardless of the reactivity of the undamaged material. The mechanism of the production of the radiation damage due to energetic ions in graphite has been studied in some detail by several groups[197–204]. It seems to be well established that the large stress observed in the surface layer after the ion bombardment[197] is due to the shrinkage of the crystal lattice in the a crystallographic direction and an accompanying expansion along the c-axis, both occuring as a result of the formation of vacancies and interstitial clusters within and between the original basal planes respectively[198, 201, 202].

Similar chemical effects on sputtering have been found also for silicon[196, 205], and silicon carbide and boron carbide[93, 94]. In the case of the carbides the surface becomes depleted in carbon as sputtering progresses at elevated temperatures, and chemical sputtering tends to disappear. Since silicon displays a pronounced maximum of reactivity with atomic hydrogen at $\sim 80\ ^\circ C$[196] an opposite effect may arise at low temperatures.

Apart from the case of hydrogen on carbon and carbides, further chemical reactions and chemical sputtering have been reported[95, 147, 196] and one might anticipate that for situations where a volatile compound may be formed, chemical sputtering or chemical evaporation due to thermal H-atoms reactions is a possible mechanism in some temperature range[97]. For example, oxides subjected to bombardment by hydrogen may lead to formation of H_2O as well as surface hydroxyl groups[98]. Certain other hydrides may be volatile.

Data are required in this area for all candidate materials that contain atoms from the right hand portion of the Periodic Table on reaction rates as a function of particle energy, wall temperature, wall composition, surface composition and production of gaseous products as a result of chemical reaction.

The most pressing need is for data on stainless steel and the insulators used in present-day machines, especially in light of the fact that these materials are the first choice from an engineering standpoint for all devices. The general requirement is that sputtering of all candidate materials be studied for hydrogen, C and O impact; this must be performed up to temperatures of 1000 °C. Any marked deviations from expected physical sputtering behavior at elevated temperatures would suggest chemical sputtering, which would then require explanation from a chemical point of view.

5.3 Desorption

Desorption of impurities adsorbed on the first wall and limiters may significantly contribute to the total impurity influx into the plasma during a discharge. In addition to thermal desorption, one must consider ion and neutral atom impact desorption, electron induced desorption and photon induced desorption. Since thermal desorption is a relatively well understood process it will not be further discussed here although it is obviously of great importance to Tokamak operation.

5.3.1 Photon and Electron Induced Desorption

It is now known that the quantum process of photodesorption does occur and that it is surface-material sensitive. For clean metals it is an extremely inefficient process if it occurs at all[99–101]. Quantum efficiencies appear to be of the order of $< 10^{-9}$ molecules/photon, with corresponding cross sections of the order of $< 10^{-27}$ m^2. Photodesorption from semiconductors appears to be a very efficient process with quantum efficiencies approaching 10^{-2} molecules/photon and cross sections as high as 10^{-21} m^2. Photodesorption is material dependent and therefore the material must be specified as well as the photon flux as a function of wavelength before estimates of the contribution to overall desorption processes can be made. Additionally the gaseous species present in the system which may adsorb and contribute to photodesorption must also be known.

Mass analysis of the desorbing species indicates that CO_2 is invariably a major component regardless of the substrate involved[102]. Carbon, which can be present as a surface contaminant, can be oxidized by gaseous oxygen and a photon induced mechanism can lead to the breaking of the chemisorption bond, leading to release of gaseous CO_2. X-rays of 15–50 keV, as well as bremsstrahlung radiation[104], have also been shown to induce photodesorption[103].

As compared to photodesorption, electron-induced desorption, E.I.D., is relatively independent of substrate material. The electrons interact directly with the absorbed gases, being only slightly affected by the underlying substrate. The impinging electrons induce transitions in the adsorbed molecule, some of which occur to dissociative states giving rise to desorbing fragments which have attained sufficient excess kinetic energy to be able to leave the surface. As energy levels of the adsorbed molecules are not drastically changed compared to the gaseous species the cross sections for electron interaction should be quite close to those for gas phase electron-molecule collisions. Thus, cross sections for production of desorbing ion fragments range from a maximum of 10^{-22} m^2, and indirectly determined cross sections for neutral desorption range from a maximum of approximately 10^{-20} m^2 [101].

In the area of photon-induced desorption the following additional studies are required:

1) more systems must be studied with emphasis on carbon and metal oxides;

2) measurements of the wavelength dependence must be extended from the near U.V. (~ 2000–4000 Å) to the extreme U.V. and x-ray (~ 10^{-3}–10^{-1} Å) regions of the electromagnetic spectrum;

3) theoretical studies are needed to establish a more complete understanding of the process so that predictions can be made concerning desorption rates and the effect of various parameters on the photodesorption process.

In the area of electron induced desorption, information exists on some simple systems (usually pure metals) using generally monochromatic energy electrons (~ 100 eV), but there are order of magnitude discrepancies between published values. The required additional data include:

1) cross section vs. electron energy over a large energy range (e.g., 0–10 keV) for systems of interest;

2) effect of surface temperature on cross section values;

3) nature of desorbing species (ratio of neutrals to charged particles);

4) energy distribution of desorbed species; and

5) theoretical modeling of the basic processes.

5.3.2 Ion and Neutral Impact Desorption

Ion and neutral impact desorption occurs from the sputtering of adsorbed species and could be a major mechanism by which impurity atoms are ejected from the wall surface. If the cross sections for desorption of surface species are large, ion and neutral impact desorption may occur more readily than the sputtering of wall material.

Winters and Sigmund[105] have investigated the sputtering of chemisorbed nitrogen from tungsten using flash filament techniques. The high sputtering yields for the inert gases were accounted for by theoretical estimates which considered binary collision processes. Another experimental technique is to monitor the remaining surface coverage by a surface analytical method. Low energy ion scattering has been used by Taglauer, et al.[106] for such studies.

Factors influencing the desorption cross sections of surface species are:

1) the composition, binding energy, and location of the adsorbate;

2) the composition, topography, and temperature of the substrate; and

3) the composition, energy, and direction of the ion or neutral flux at the wall surface.

Since desorption yields are strongly dependent on the energy of the incident particle, it is necessary to compare data with incident particle energies that are similar to those expected in the particle outflux from the plasma. Only limited data are available for desorption rates of interest to Tokamak operations. Ion impact desorption data have been obtained by McCracken[107] using 5.6 keV D^+.

Desorption yields for the three above mentioned processes can be combined with the expected particle fluxes to give estimates of the impurity influx for these desorption processes. The corresponding impurity influx from deuterium ion impact desorption is estimated to be about two orders of magnitude larger than that expected from either electron or photon induced desorption. Therefore, the impurities released into the plasma during a short discharge (~ 1 s) will be dominated by deuterium impact desorption while electron and photon induced desorption will occur only to a much more limited extent.

5.4 Vaporization

Thermal vaporization of wall, limiter, and beam dump material could be a serious contributor to plasma impurity release. Relevant data on the vapor pressures of candidate first wall and limiter materials are in general known. It might be thought that adjusting the machine operating parameters so that the equilibrium wall temperatures are below a predetermined upper limit would keep the plasma impurity release rates due to vaporization processes within acceptable limits, particularly if materials with low vapor pressures are used. However, the situation is complicated by several phenomena which may give erroneous vaporization rates if equilibrium vapor pressure data, which are generally obtained from Knudsen cell measurements, are used as a basis for the estimates. Changes in surface topography and surface roughness due to sputtering, blistering and gas re-emission will greatly alter the kinetics of the vaporization process from those obtained using equilibrium techniques. The high power levels of photon radiation emanating from the plasma and its partial absorption by the near-surface regions of the wall is expected to raise the surface temperatures to levels which depend not only on the material, but also on its surface condition which may vary in different locations in the Tokamak. Finally, coatings are being seriously considered as a method of achieving desirable surface properties on acceptable but different bulk substrate materials. The bonding character of coatings can result in serious alterations of evaporation temperature. The effects of these phenomena on vaporization rates under realistic machine conditions are areas of work in which a greatly expanded effort needs to be made before data useful to machine designers can be generated.

5.5 Blistering and Flaking

Energy transferred to the bulk and near-surface region by impinging ions damages the lattice structure by creating vacancies, defect clusters, etc., and introducing the ion species as substitutional and interstitial impurities. Given sufficient thermal energy, which is readily available in the fusion reactor, the implanted impurities will migrate to grain boundaries and hence to the surface, to be re-emitted. Hydrogen, being extremely mobile, is readily re-emitted. Helium mobilities are many orders of magnitude smaller than those of hydrogen in many solids (metals in particular) so helium will more readily aggregate into bubbles which migrate via stress and thermal gradients to form layer bubbles. The result is that either the bulk material will swell or the bubbles will appear as blisters on the surface to relieve the stress[108)]. In contrast, helium is more easily desorbed from graphite than hydrogen and the radiation damage is due to structural changes of the lattice rather than helium bubble formation[84, 197)]. Rupture and exfoliation can follow which will result in impurities entering the plasma. Erosion rates associated with helium blister exfoliation may exceed the sputtering produced by the helium atoms escaping from the plasma.

Hydrogenic ion-induced blistering has been observed, but by itself will not likely present a problem to the first wall at elevated reactor operating temperatures. The effects of blistering at high primary fluxes such as those found at beam dumps and components of neutral beam injectors, divertor throats and bombardment plates, are not known.

The simultaneous irradiation with hydrogenic and helium ions can also present a serious blistering problem because of synergistic effects. Although there is a paucity of data, a D-T fusion environment is highly conducive to such synergistic effects. Such an environment contains 3.5 MeV He^4 formed by D-T reactions and 1 MeV T, 3 MeV H, and 0.8 MeV He^3 formed by the D-D reactions in advanced fuel reactors. The energy spectrum of the central plasma can be expected to be Maxwellian at the Tokamak operating temperature (~ 10 keV). There is little or no information on the interaction between blistering and externally induced stress. Thermal and mechanically induced stress will influence diffusion and this will have a direct effect on blistering.

Blistering has been couched traditionally in the context of materials performance in a fusion environment, and, as such, has been considered a problem in the long-term sense. There exists, however, some indication that blistering-like phenomena may occur when hydrogen discharge cleaning techniques are used to condition the plasma containment chamber walls of existing devices. (See Sect. 5.6.2.)

As discussed above, blistering will also occur at the divertor throat and particle bombardment plates, and will therefore be important for future reactors employing these devices. Blistering of neutral beam injector components must also be considered and therefore blistering should be considered a potential problem in a broader sense than has traditionally been accepted.

Data have been collected in recent years on the effects of He implantation in metals[109–111)]. Gas re-emission measurements have shown that, at low doses, essentially all of the helium is retained in the lattice. At a critical dose, which depends on a number of variables (energy, temperature and material) the surface deforms and gas re-emission increases dramatically. This phenomenon has been observed in a large number of metals and alloys.

Surface deformation effects have been found to be significantly altered with multiple energy implantation[112, 113]. In one experiment[112] multiple energy pre-implantations were made to create a uniform helium concentration extending from the surface to a depth of ~ 700 Å. During subsequent 20 keV implantations to fluences beyond the critical dose it was found that surface deformation was virtually eliminated when the preimplanted helium concentration was $\gtrsim 0.3$ atom fraction.

The helium implantation profiles, after incorporating the angular distribution for the 3.5 MeV alpha particles envisioned in a power reactor, are roughly Gaussian in shape and peaked at 1.5 to 4 μm. Calculated helium profiles show that surface deformation can occur if a critical concentration is reached. The likelihood of blistering is then determined by a competition between the accumulation of helium at a depth of several microns, and the erosion of the surface by D-T sputtering.

Any redeposition of sputtered wall atoms between burn cycles will make blistering more probable and more rapid. It appears that stainless steel first walls will suffer exfoliation for plasma edge temperatures below 100 eV, at all first-wall temperatures. Blistering can be expected to occur within ~ 10^3 h of operation of a fusion power reactor.

5.6 Other Mechanisms

5.6.1 Unipolar Arcing

Although observed quite early in the development of plasma devices, unipolar arcing has only recently gained widespread attention as both a major source of plasma impurities and as a potentially severe erosive effect for first walls, limiters and other material surfaces in close proximity to Tokamak plasmas.

The principal method of introducing metal impurities into early pinch discharges was considered to be arcing. The externally applied voltages, although low, still permit the occurrence of unipolar arcing between the plasma and the wall when driven by the sheath potential. Local electron emission from a cathode spot is balanced by a uniform flow back to the surface of energetic electrons in the tail of the Maxwellian distribution.

It has been shown that under normal discharge conditions in many of the present Tokamaks the radiated power from metal impurities dominates the power balance. The metals presently observed are predominantly iron from the stainless steel walls, and molybdenum or tungsten from the limiters. When the torus has been exposed to the atmosphere, oxygen and carbon are initially important. After a short period of operation and the use of hydrogen discharge cleaning (to be discussed in Sect. 6.2.), the low-Z impurities are reduced and the metal impurities are responsible for radiating more than 50% of the ohmic power. The evidence for the nature and concentration of the impurities comes from vacuum UV spectroscopy, from the radial distribution of the total radiation, and from x-ray diagnostics.

Cohen, et. al., have discussed the possible mechanisms of impurity release in PLT[81]. Hydrogen sputtering has been unable to explain the observed metal fluxes.

There is still the possibility that the metals may be sputtered by multiply charged low-Z impurities such as oxygen and carbon. Self sputtering is also a possibility but at present the question is still to be resolved. Evaporation as a possible source of metal impurities has also been dismissed. Unipolar arcing appeared to be the most likely mechanism leading to the observed impurity level[81)].

By thorough investigation of the surface of the Tokamak torus, a great deal of evidence for the presence of arcs has been found on the torus wall, on the fixed limiters, on movable limiters and on specially prepared probes which have been inserted into the vacuum vessel[81, 114)]. Two types of arcs have been observed; one is fernlike, the other consists of narrow linear tracks. The material removed from the surface during an arc is ejected in the form of molten blobs as well as in atomic form. The duration of vacuum arcs is typically 10^{-5} to 10^{-2} s.

Although a great deal of work has been done in the general field of gaseous discharges, it is clear that the severity of the unipolar arcing effect necessitates its specific study in experimental plasma devices. In addition, standardized techniques for producing arcs and analyzing their behavior should be devised. These techniques could serve as a means of screening candidate fusion materials for arcing, as well as allowing for the analysis of the means by which the consequences of arcing may be minimized. One possible technique involves the use of high-power plasma devices. Preliminary results indicate that arcing is, in fact, a function of metallurgical variables such as alloy morphology, impurity level and distribution, and surface preparation. In fact, the minimization of the effects of arcing may be possible through the control of such material variables as well as through the control of the near-surface plasma condition (i.e., cool plasma gas blanket). Unipolar arcs can be reduced by lowering either the electron temperature or density, but since the theoretical threshold is at $T_e \gtrsim 3$ eV it will be difficult to eliminate completely. The use of hydrogen gas puffing, i. e., introducing an additional burst of gas into the torus during a discharge, has been effective both in reducing the effective Z of the plasma ($Z_{plasma} = n_e/n_i$) and in decreasing the depth of the observed arc tracks. This probably arises due to cooling the plasma boundary. The use of divertors would also be expected to be effective both by reducing the plasma density near the walls as well as by screening impurities entering the discharge from the wall. The use of divertors will be discussed in greater detail in Sect. 6.1.

5.6.2 Metal Snow

During a plasma discharge, the walls of toroidal devices are exposed to intense fluxes of atomic ions and neutral particles which escape from the plasma and strike the first wall leading to a buildup of hydrogen in the near surface regions as discussed in Sect. 4.2. It has been suggested that as a consequence of the limited mechanical strength of the wall material pockets of trapped hydrogen gas lying closest to the surface may "explode" producing a "metal snow"[115)]. The bursting of a pocket can lead to the following consequences:

1) sharp local pressure increases in the confinement vessel;

2) release of metallic dust, i. e. intergranular material, blown into the pressure vessel;

3) deeper lying pockets can become unstable and "explode" in an avalanche-like effect;

4) the effective area of the experimental vessel is increased by an amount equal to the surface of the now opened crevice.

However, this pocket "explosion" effect appears not to be an annoying long lasting nuisance, but presents only a transient character: Prolonged exposure to intense atomic particle fluxes leads to its disappearance[115]. The presently available experimental data do not allow a final conclusion about the significance of this suggested mechanism of the plasma contamination to be drawn and more study is necessary.

5.7 Interactive Effects

Single-component irradiations have provided useful information concerning certain projectile/surface interactions which contribute to plasma contaminant release and surface damage and erosion, but they cannot yield any information about interactive (synergistic) surface effects. These arise when two or more plasma radiation components interact simulaneously with surfaces, together producing plasma contaminant release and/or surface erosion either larger or smaller than that expected from a simple summation of the effects caused by the individual radiation components. Interactive effects can also occur after sequential irradiations.

The present knowledge of the influence of synergistic effects on plasma contaminant release, surface damage and erosion is extremely limited. Furthermore, in existing plasma devices only integral surface effects, resulting from the simultaneous interaction of radiation components with surfaces, have been observed. Such integral effects cannot readily be extrapolated to plasma devices of different design and operating parameters. The present paucity of information on interactive effects allows one only to speculate in a very cursory way about their potential influence on plasma contaminant release and surface erosion in fusion devices. Since the parameters of those plasma radiation components which interact with the first wall are expected to be significantly different from those parameters typical for neutral beam injector dumps (and possibly for limiters), a distinction between the effects expected for such components will be made wherever it seems appropriate.

In addition to thermal desorption, gas desorption has been found to result from electron, ion and photon bombardment of surfaces. Therefore, simultaneous particle and photon bombardments can be expected to alter desorption rates, as well as the nature and charge distribution of the desorbed species. Furthermore, simultaneous bombardment of a surface by neutrons and ions could affect diffusion processes, e.g., by radiation-induced segregation. In turn, desorption processes can be influenced by altering the diffusion of species from the bulk to the surface. The type, energy, and angular distribution of particles expected to strike neutral beam injector dump areas (such areas can represent 1/9 of total first wall area) can cause synergistic effects on gas desorption which can be quite different from those expected from the interaction of plasma radiations with the first wall.

The simultaneous bombardment of a metal surface by energetic and chemically-active deuterons and by energetic but chemically inert helium ions can be expected to result in interactive chemical and physical trapping, and gas release. The simultaneous formation of metal deuterides and helium blisters could lead to altered plasma contaminant release. Again, synergistic effects on gas release and vaporization from beam dump areas can be expected to be quite different from irradiated first walls, due to the significant difference in the parameters and types of the radiation components.

Chemical, and possibly also physical, sputtering rates can be affected by interactive effects, due to the interaction of chemically active and inert particles with surfaces. Changes in the surface topology (e.g., roughness and stresses) and the chemical state may affect not only sputtering rates, but also the species of sputtered particles and their energy and angular distributions. Again, the effects on beam dump areas may be quite different from those on the first wall or the limiter.

Synergistic effects are also expected to affect surface damage and erosion in fusion devices. For example, low-energy photons of sufficient flux, striking the skin of blisters produced simultaneously by energetic particle bombardment, can cause skin heating to a temperature above that of the substrate due to reduced thermal contact of the blister skin with the bulk. In turn, this may reduce the yield strength of the skin and cause accelerated blister growth, rupture, and possibly flaking. Interactive effects on surface roughening and erosion can be expected from the simultaneous occurrence of physical and chemical sputtering of surfaces bombarded simultaneously by chemically active and inert particles. The power deposition by the simultaneous impact of ions, electrons, and photons on blistered surfaces may lead to synergistic effects on vaporization. The simultaneous formation of deuterides of the target material and helium blisters could lead to interactive effects on the changes of the chemical state of irradiated surfaces. Furthermore, the simultaneous bombardment of a surface by ions and neutrons can lead to synergistic effects on irradiation-induced surface segregation. It is very likely that the types of interactive effects which affect processes leading to surface damage and erosion will be different for the neutral beam injector dump areas (high deuterium radiation component) and the first wall (comparatively low deuterium radiation component).

Plasma simulation experiments along with single and simultaneous multiple ion irradiation studies are needed to investigate the possibilities of interactive effects[116].

6 Impurity Control

The previous discussion has shown that plasma impurities present a complex set of problems whose solution is crucial to the successful operation of fusion reactors. The many and often subtle factors that govern plasma-materials interactions are still only partially understood. Consequently, the methods used today to control impurities are to a degree empirical in nature and cannot yet be precisely defined. It is likely that in the end a variety of approaches will be used to keep plasma impurities at minimal levels. The techniques are conveniently divided into divertor and non-divertor methods. The latter depend on modifications of one sort or another of the composition or structure of the surfaces facing the plasma. As will be seen in Sect. 6.5., methods for enhancing the ion fraction sputtering yield of surfaces can be expected to make their greatest contribution to impurity control in divertor operated machines. Therefore, divertor and non-divertor methods are to a degree interrelated.

6.1 Divertors

The magnetic divertor was already mentioned with reference to the Tokamak power reactor design, UWMAK III (Sect. 2.2 and Fig. 7). The objective of a magnetic divertor, first suggested by Spitzer[117)], is to reduce the plasma wall interaction by magnetically channeling charged particles near the wall into a remote divertor chamber where they are trapped and/or pumped away. Thus, the primary fluxes of charged particles towards the wall are reduced (unload mode) as are also the fluxes from the wall into the plasma (shielding mode). The fast charge exchange neutrals and sputtered neutral species are not affected by the divertor.

In order to reduce significantly the impurity influx into the plasma they have to leave the wall as ions (See Sect. 6.5.) or be ionized within the scrape-off layer before reaching the separatrix (see Fig. 7). The latter denotes the magnetic surface separating the magnetic lines which are closed within the confinement region and those which penetrate into the divertor chambers (Figs. 7 and 19).

There are three fundamental categories of divertors: toroidal, poloidal and bundle divertors. The toroidal divertor has been tested on stellarators but it is unsuitable for Tokamaks because of its geometric incompatibility[118)]. Figures 19a, 19b show schematically a poloidal and a bundle divertor respectively. The bundle divertor has been tested at Tokamak DITE[119)] since 1976; examples of devices utilizing a poloidal divertor are DIVA (JAERI, Japan, operation since 1974), T-12 (Moscow, USSR, 1977), ASDEX (Garching, F.R.G., 1978), PDX (PPL, 1978). The article by Shimomura and Maeda[120)] summarizes the state of art in 1978. A strong reduction of plasma core radiation by heavy ions has been demonstrated by all divertor exper-

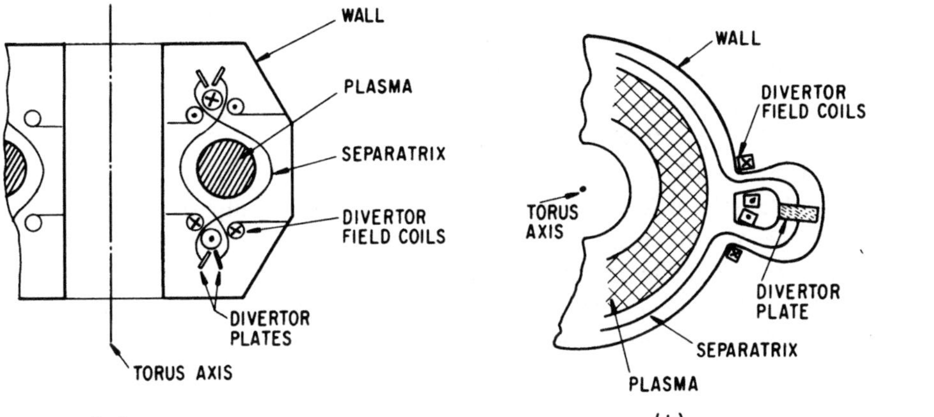

Fig. 19. A schematic view of *a*-poloidal and *b*-bundle divertor (see text)

iments performed so far. A decrease of the level of low-Z impurities, such as oxygen was seen in DIVA but it was not verified in DITE. Experiments with impurity injection into the main as well as into the divertor chambers have shown that the back diffusion from the divertor is low.

Since the plasma is in part pumped away by the divertor, efficient refueling is necessary in order to keep a constant plasma density. An adequate refueling method has to maintain a sufficiently low rate of production of charge exchange neutrals. The possible effects of the divertor on the MHD-stability of Tokamak plasmas and a detailed understanding of the processes in the scrape-off layer require further investigations.

6.2 Discharge Cleaning

Discharge cleaning has become a standard method of surface conditioning in present Tokamaks because it helps to reduce oxygen and carbon impurities to a tolerable level. Discharge cleaning has been a well known method used by physicists studying glow discharges and has also been successfully applied to clean the stainless steel wall of the Intersecting Storage Rings (ISR) at CERN[121)]. The vacuum requirements are extreme in the latter case since the accelerated particles have to travel distances of many thousands of kilometers before undergoing collisions in the storage rings. After 24 h of baking at 300 °C a base pressure of $\sim 10^{-11}$ mbar had been reached, which was increased orders of magnitude during the operation because of impurities desorbed from the wall due to bombardment by ions and photons. Since such a pressure increase prevents effective operation of the ring, a detailed investigation of various surface cleaning procedures has been performed by Mathewson and co-workers[121)]. It turned out that only discharge sputter cleaning provided a sufficiently clean surface.

The recent progress in the application of discharge cleaning techniques to Tokamaks is apparent from a comparison of the papers presented at the conferences on plasma wall interaction during the last ~3 years[38–40)]. It is generally accepted that a relatively cold hydrogen discharge plasma is the most appropriate method compatible with Tokamaks, although various groups utilize somewhat different discharge techniques, such as 50 Hz discharge[122, 123)], 5 kHz discharge[124)], rf discharge and thermally produced H-atoms[125, 126)], and a weak, pulsed toroidal discharge[52, 127)]. The latter method seems to be preferred since it requires no equipment in addition to that provided by the Tokamak itself. The discharge duration varies between ~0.5 ms[52)] and 20 ms[127)] with a repetition frequency of $\sim 120\ s^{-1}$ and $\sim 3\ s^{-1}$ respectively. The toroidal field is low and the electron temperature is typically $\lesssim 30$ eV. In order to obtain the efficient removal of oxygen and carbon impurities the temperature of the vacuum chamber has to be raised to ~ 200 °C. The optimal cleaning conditions are found by following mass spectroscopically the rate of production of H_2O, CH_4, CO and CO_2. According to recent results on TFR the initial coverage of the wall surface with a monolayer of oxygen was reduced to less than 0.01 monolayer after 2 h of such a discharge cleaning procedure. Similar results are obtained by the Taylor method[124)].

As mentioned at the beginning, the present discharge cleaning techniques provide Tokamaks almost free of low-Z impurities. However, very little is known about the mechanisms of the surface chemical processes[128, 129] that occur during discharge cleaning and it is possible that such knowledge could help to develop even more efficient and economical cleaning techniques.

6.3 Protective Coatings

The material for the first wall has to meet a number of requirements. Besides having a low erosion rate and resulting low plasma contamination, the criteria as listed in Table 3 for selecting wall materials in the near term experimental fusion reactors have also to be considered[28]. With respect to a long term commercial power reactor the priority list of these criteria is different and the availability of the material resources has to be considered as well, as shown in Table 4.

Obviously, no one material meets all the requirements, although titanium and vanadium alloys have many desirable characteristics. A comparison of the data given in Table 4 with the priority list of Table 3 reveals, however, that stainless steel remains the primary choice for the near term experimental fusion reactors because of the existing industrial capability and a large data base.

The overal requirements on a single material can be significantly relaxed by combining several materials in the first wall design[130–133]. One alternative is to use a composite structure consisting of a metallic vacuum wall with an adherent protective coating that faces the plasma. The other choice may be a metallic vacuum wall with a nonmetallic liner placed between the vacuum wall and the plasma, such as the carbon curtain in UWMAK II and III (see Fig. 7), or a coated graphite or silicon carbide liner.

Both alternatives have several advantages in common:

1) separation of structural and surface requirements;

2) greater freedom in the material selection for the surface facing the plasma;

3) possibility to obtain desired microstructures which minimize erosion due to blistering and flaking.

In addition, the thin, low-Z coating concept offers several advantages as compared with the separate liner:

1) cooling is provided by thermal contact with the metallic wall;

2) no problem with mechanical support exists provided there is good adherence between the coating and the wall;

Table 3. Criteria for selecting wall materials in near term experimental fusion reactors[28]

1. Industrial capability and existing data base
2. Compatibility with coolants and tritium
3. Fabricability and joining
4. Mechanical and thermal properties
5. Induced radioactivity
6. Cost
7. Radiation damage

Table 4. Criteria for selecting first wall materials[a] in fusion reactors in general priority order[28)]

Criteria	Favored materials	Less favored
1. Radiation damage and lifetime		
a) Swelling (dim. stability)	Ti, V, Mo, SS[c]	Nb, Al, C
b) Embrittlement	C, Nb, V, Ti, SS	Mo, Al
c) Surface properties	V, Ti, Al, C	SS, Nb, Mo
2. Compatibility with coolants and tritium		
a) Lithium	Ti, V, Nb, Mo, SS	(Al, C)[b]
b) Helium	SS, Ti, Mo, Al, C	(Nb, V)[b]
c) Water	SS, Al, Ti	(C)[b]
d) Tritium	Mo, Al, SS	Ti, V, Nb, C
3. Mechanical and thermal properties (irradiated)		
a) Yield strength	Mo, Nb, V, Ti, SS	Al, C
b) Fracture toughness	SS, Ti, Al	V, Nb, Mo, C
c) Creep strength	Mo, V, Ti, SS	C, Al, Nb
d) Thermal stress parameter $M = \frac{2\sigma_y k(1-\nu)}{\alpha E}$	Mo, Al, Nb, V	Ti, SS, C
4. Fabricability and joining	SS, Al, Ti	Nb, V, Mo, C
5. Industrial capability and data base	SS, Al, Ti, C	Mo, Nb, V
6. Cost	C, Al, SS, Ti	Mo, Nb, V
7. Long lived induced radioactivity	V, C, Ti, Al	SS, Nb, Mo
8. Resource availability (USA)	C, Ti, Mo, Al, SS	Nb, V

[a] Alloys Ti-6Al-4V, V-20Ti, TZM, Nb-1Zr, 316SS, Al-6061. This is an illustrative list.
[b] Materials in parenthesis are unacceptable with stated coolant.
[c] *S*tainless *S*teel.

3) because of the small thickness of the coating ($\lesssim 250\ \mu m$) exotic and relatively expensive materials can be used;

4) a great variety of existing coating technologies presently exist;

5) materials with a relatively high rate of helium production under neutron irradiation can be used since the helium can readily diffuse out of the thin coating;

6) there exists the possibility of in-situ redeposition which can significantly reduce the downtime of the power plant.

Two aspects concerning the preparation and properties of protective coatings as related to controlled fusion devices should be noted. In present and near term devices, erosion of the first wall is small as compared with that of the limiter. Thus, refractory coatings on the limiter resistant against thermal shocks are desirable in the near future[130–133)]. On the other hand, the erosion of the first wall will become more improtant in future thermonuclear power reactors which have to be designed for long term operation. Remote repair or in-situ re-deposition of coatings will be required at this stage.

We shall consider the two alternative approaches in the following parts of this section.

6.3.1 Coated Limiters and Beam Dumps

As already pointed out, protective coatings on limiters may be desirable for present and near term devices such as PLT and TFTR. Similar requirements may also exist for protective plates and beam dumps in TFTR. Since these devices are not designed for continuous, long term operation with a D-T plasma, the protective coatings can be prepared by conventional techniques before installation of the limiter and beam dumps into the machine.

Among the possible candidates, graphite, tungsten, molybdenum and copper may serve as substrate materials with carbides (e. g., SiC and B_4C), borides (e. g., TiB_2) and silicides as candidate coatings. Conventional techniques such as high temperature *C*hemical *V*apour *D*eposition (CVD) and plasma spraying could be used to apply the coatings. Their use in TFTR and other plasma devices will depend on the outcome of a variety of materials testing procedures.

6.3.2 In-situ Deposition

The possible methods of in-situ deposition have been discussed by Stacey, et al.[130–132], Veprek, et al.[134], Chin and Ohkawa[135], and more recently at the "First Wall Coating Workshop"[136]. An optimal method has to be as simple as possible, compatible with the Tokamak design, and should provide ~ 100 to $200\ \mu m$ thick coatings with desirable properties within a reasonable time.

Good adherance of the coatings to the substrate is most critical with respect to thermal shocks, to radiation damage, and to chemical changes over long-term exposure to a hydrogen plasma. To achieve the desired adherence properties requires a proper choice of materials with respect to their thermal expansion and chemical compatibility, as well as careful control of the deposition conditions[137].

Obviously all methods involving corrosive reactants and/or products such as halides are unsuitable for in-situ deposition in Tokamaks because of the danger of undesirable chemical side reactions with reactor components. Another limitation is the wall temperature required for the particular deposition method, since it is an extremely difficult technological task to heat large Tokamaks to ~ 500 °C and achieve a uniform temperature over the whole vacuum vessel[138]. Most of the conventional CDV methods utilize much higher temperatures which have to be well controlled within small limits[135, 137]. Thus, the presently available conventional CVD techniques do not meet the requirements for in-situ deposition and new heterogeneous systems must be developed if this approach is to present a feasible alternative. Such systems have to operate at temperatures below ~ 500 °C and should involve non-corrosive reactants and products.

There has been great progress in the field of *plasma spraying* during the last 10–20 years and it is today a well developed technique currently being used in industry[139, 140]. Various materials ranging from refractory metals, oxides, nitrides and carbides to organic polymers can be deposited in this way. The apparent limitation is the requirement of having a good match of the thermal properties of the substrate and the coating. The achievable deposition rate allows preparation of a well adherent coating over an area of ~ 20 m^2 within an hour. Thus, ~ 100 h would be necessary to deposit a protective coating on the wall of a Tokamak reactor of the size of UWMAK III. Simultaneous operation of several plasma spraying devices may reduce the time requirements.

The obvious problems associated with the application of plasma spraying in a Tokamak power reactor arise from the necessity of using remote handling techniques because of the induced radioactivity. Such techniques for large scale plasma spraying in geometrically complicated devices are not presently available and would have to be developed.

The material of primary interest suggested for the first wall coating by Stacey et al. was beryllium[130–132]. The alternatives are beryllium oxide, boron carbide, boron nitride and elemental boron[133]. Because of the pronounced chemical attack of pure elemental boron by a hydrogen plasma observed by Kingcade et al.[141], (see also Ref.[142]) boron rich borides of the overall composition $B_{48}Me_x$ with $x \lesssim 4$, Me = C,N, and various metals, could be more suitable[142].

The application of *low pressure plasma* CVD for in-situ redeposition was suggested by Veprek, et al.[134] and more recently by Chin and Okhawa[135] and by Rovner, et al.[143]. This technique was successfully applied to the depostion of various dielectric materials for the electronics industry[144], for optical fibres[145], and for deposition of amorphous semiconductors[146].

With respect to the re-deposition of a coating on the first wall, the obvious advantage is the compatibility with Tokamak operation since it requires a minimum of auxiliary equipment and modification, particularly if a continuous, high frequency discharge is used. The power requirements for such an approach appear to be acceptable[134]. An rf heating system, if incorporated into the reactor design (e. g. UWMAK III) could be advantageously used for these purposes.

In order to obtain the excellent dielectric properties desirable for electronic applications the growth rate of coatings is limited to about 10 μm per hour at ~300 °C, but well adherent layers can be grown at much higher rates if the requirements on dielectric strength are relaxed and temperatures of ~ 400–500 °C allowed[147]. Since the depostion takes place simultaneously on the whole surface, only several hours should be necessary to redeposit a ~100–200 μm thick coating[134].

In contrast to plasma spraying, low pressure plasma CVD does not require any remote handling technique. However, there is yet no experience with large scale applications of this method, particularly in metallic vessels. This is the first goal towards which future studies have to be directed. Very little is also known about the adherence of the coatings deposited by low pressure plasma CVD and their resistance against thermal shock. The choice of the best material for such coatings is presently open to discussion.

Carbon, although advocated by some researchers[143], does not appear to be acceptable as a coating because of its high chemical sputtering and its reactivity with atomic hydrogen at the expected temperature of the first wall of 500–700 °C. Bulk graphite may, however, prove to be suitable as a mechanical support of a separate, appropriately coated, liner and/or limiter.

The application of silicon carbide is not without obvious problems because of its high sputtering yield[148] and the difficulties associated with depositing a well bonded, homogeneous compound. Anderson and Spear[149] have found that "silicon carbide" prepared by this technique contains few Si-C bonds and it represents rather an agglomerate of silicon and graphite. A more detailed study of plasma deposited hydrogenated a-$Si_{1-x}C_x$ films has been performed by Wieder et al.[211]

Because of its lower sputtering yield boron carbide might appear more suitable[148] provided the thin film coating would show better resistance against thermal shocks than bulk boron carbide[55]. However, there is limited experience with the deposition of this material at low temperatures and similar problems as in the case of SiC are to be expected. The mechanical and thermal properties of such materials deposited at low temperatures might be significantly improved by using ion plating[150], but it is not clear yet if this technique can be applied to large scale devices. This technique would also require some kind of remote handling similar to plasma spraying, and achieving a reasonable overall redeposition rate might represent a serious limitation. Nevertheless, similar effects can be obtained by biasing the R. F. plasma with respect to the vessel wall as in the case of low pressure plasma CVD discussed above.

Many of the material problems mentioned can be circumvented by application of boron rich borides as suggested recently[142, 151]. In particular, the impurity stabilized α-tetragonal polymorphs as represented for example by $B_{48}B_2C_2$, $B_{48}B_2N_2$, and $B_{48}B_2Me_2$ appear promising. These are refractory materials which can accomodate relatively large deviations of chemical composition from the exact theoretical stoichiometry without significantly altering their properties, since impurities can be substituted by isolated boron atoms. Encouraging results on the deposition of boron and carbon doped boron coatings at $\lesssim 400$ °C were obtained recently[142].

6.4 Trapping Surfaces

In Sect. 4.2. on gas re-emission and trapping, the beneficial effect of titanium gettering on plasma recycling was discussed. The results of gas re-emission measurements for several active metals are shown in Fig. 16. Because of the importance of trapping of hydrogen isotopes both for plasma recycling and for divertor action, the characteristics of trapping surfaces need to be understood in detail. The subject is of great importance to impurity control since sputtering, which may become the dominant high-Z impurity release mechanism in future high temperature Tokamaks, is closely linked to plasma recycling.

Various schemes for attenuating the energetic charged particles in the divertor region have been mentioned, including direct conversion, glancing collisions with surfaces, collisions with cold gas, reactive (chemical) and non-reactive trapping. Chemical trapping occurs when energetic particles, in particular H^+, D^+ and T^+, interact with certain metals to form hydrides, deuterides and tritides. The particles are captured by the metal and quickly come to thermal equilibrium. In the case of He^+ particles, no chemical trapping can occur but a certain fraction of incident particles will be retained by implantation in the metal. Only titanium coatings prepared by thermal, in-situ evaporation have been used in Tokamaks so far. In view of the high temperatures necessary for hydrogen desorption from titanium (Sect. 4.2.), other materials are desirable for devices operating with D-T fuel.

6.4.1 Mechanism of Chemical Trapping of Deuterium and Tritium in Metals

Experimental measurements on deuterium trapping in solid metal targets[5, 152–156] have shown that this may be a promising technique for "pumping" a useful fraction of the ion flux in the divertor.

The total particle flux from a ~ 6 GW (thermal) reactor will be $\sim 6 \times 10^{22}$ ions s^{-1}. To maintain a sufficiently large mean free path in the divertor region for the incoming ions (say 10^{-5} mbar pressure) would require a pumping speed of $\sim 10^8$ ℓs^{-1} at thermal energies. Handling the required throughput of D, T and He by conventional pumps would pose formidable problems. Chemical trapping could reduce pumping speed requirements by a factor of about 20 assuming 100% trapping efficiency, 5% burnup and zero trapping of He. A great incentive therefore exists for investigating trapping mechanisms and efficiencies.

The trapped D and T could be recovered by heating the trapping surface outside the reactor to temperatures at which decomposition pressures of 5×10^{-2} mbar are produced. Vapor booster pumps handling $\sim 10^3$ mbar ℓs^{-1} at these pressures could then be employed.

Studies of trapping in solid targets have shown that the trapping efficiency of hydrogen isotope ions in solids depends on ion energy, bombardment time and target temperature. Trapping also has been found to be most efficient in those metals which form stable metal hydrides. Some results obtained by McCracken, Jeffries and Goldsmith[153] are shown in Fig. 20 where trapping efficiency after a fixed arbitrary bombardment time (total dose of 5×10^{22} ions m^{-2} of 18 keV D^+) is plotted against temperature range for each of four metals. It is to be noted that the trapping is efficient over only a certain temperature range.

These results can be understood on the basis of the following considerations[153]. The 18 keV D^+ ions have a mean range of $\approx 2 \times 10^{-7}$ m[158]. After slowing to thermal energies, deuterium diffuses through the lattice without being able to escape from the surface over a certain temperature range because the activation energies for diffusion are much lower[159–161] than the heats of formation of the metal deuterides[162–165]. The decreasing trapping efficiencies at higher temperatures are due to the increasing probability of the diffusing D atoms to overcome the potential barrier for escape from the surface. The expected correlation between the heats of formation of the (stoichiometric) metal deuterides and the temperature at which the trapping efficiency has decreased to some arbitrary value is observed[153]

At temperatures in the liquid nitrogen range and below, the diffusion coefficient of deuterium in metals falls to very low values and therefore the deuterium concentration rises rapidly in a layer whose depth equals the projected range of the incident

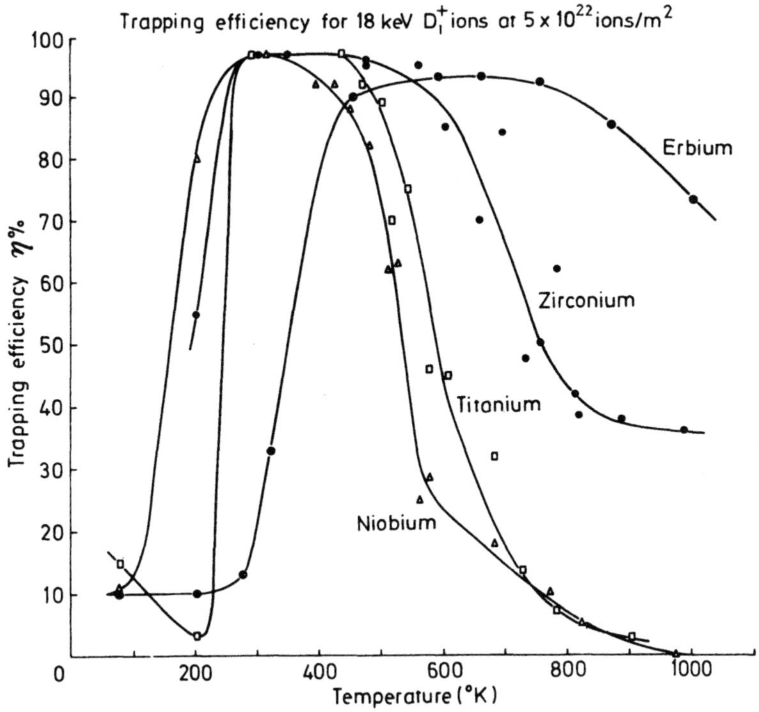

Fig. 20. Trapping efficiency of Nb, Zr and Er for 18 keV D^+ ions as a function of temperature. Total ion dose = 5×10^{22} ions m^{-2} [153)]

ions. It has been found experimentally for all the metals studied to date[153)] that at low temperatures, the trapping efficiency decreases quite abruptly at a dose equivalent to 10^{22} ions m^{-2}. If distributed evenly throughout the solid to a depth equivalent to the initial projected range of the ions in the metal, the M/D ratio would be of the order of unity. It appears that as the surface layers are converted to a bulk hydride phase, no further uptake of deuterium occurs.

At higher temperatures, the trapping efficiency remains high at doses larger than 10^{22} ions m^{-2} because the diffusion coefficient is large enough to prevent the build-up of saturation layers near and on the surface.

6.4.2 Characteristics of a Trapping Surface in a Thermonuclear Reactor

The preceding discussion has made clear that trapping is inefficient both at low temperatures (for a dose of $\gtrsim 10^{22}$ ions m^{-2}) and at high temperatures. The low temperature limit is determined by the low diffusion rate of the gas in the solid leading to saturation of the surface layers of the metal. The high temperature limit is determined by the decomposition pressure of deuterium in the metal deuteride. However, at intermediate temperatures, trapping efficiencies greater than 90% may be obtained.

For practical application in a thermonuclear reactor, it is important to determine whether the total fluence which can be trapped with high efficiency fulfills the requirements placed on a trapping surface.

The 6×10^{22} particles leaving the 6 GW(t) reactor per second carry 150 MW of power. With approximately 5×10^{1} m^2 of trapping surface (some 10% of the reactor vacuum wall area), the power loading becomes 3×10^{6} Wm^{-2}. The maximum ion current density which has to be considered is $\sim 1 \times 10^{21}$ ions $m^{-2}s^{-1}$ or 1.9×10^{5} $mA \cdot m^{-2}$. Such ion current densities incidentally are experimentally accessible with ion sources of the duoplasmatron type.

In a recent study[156] a tensimetric technique was developed for measuring 15 keV D^+ trapping in Ti metal. The formation of the TiD_2 phase in the surface and nearsurface regions was monitored by decoupling the $\theta - 2\theta$ scan of an x-ray diffractometer as well as by scanning electron microscope studies. Micro sized "reaction zones" appear to serve as nucleation centers for the TiD_2 phase. Insight into the mechanism of chemical trapping was obtained by correlating the results from the three types of measurements.

6.5 Near Surface Modifications and Secondary Ion Emission

Proposals for reducing the high-Z impurity influx fall into two categories: modifying the plasma edge and modifying the wall surfaces. Assuming that most of the influx is the result of sputtering by charge-exchange neutrals originating in the plasma edge, the sputtering yield would be reduced by lowering edge temperatures by techniques such as cold gas puffing. Alternatively, a magnetic divertor creates a "scrape off" region in which ions at the plasma edge are prevented from colliding with the wall and ions ejected from the wall are removed before they enter the plasma (see Sect. 6.1). For divertor action to be effective, neutral sputtered particles must be ionized, requiring the use of a gas layer[166] to effect charge exchange ionization. The risk inherent in this approach is that the gas layer would most likely increase the charge exchange neutral flux to the wall and thus be self-defeating.

A novel approach to impurity control first suggested by Krauss and Gruen[167] is to select wall materials in such a way as to maximize the fraction of atoms which are ejected from the wall as ions. Three possible fates of particles ejected from a Tokamak surface facing the plasma are shown in Fig. 21. If the particle ejected from the surface is neutral, it will penetrate the scrape-off region (in a divertor operated Tokamak) to a depth determined by the kinetic energy and trajectory of the particle, and density of the gas in this region. If the particle becomes ionized in the scrape-off region, it will be swept into the divertor. Otherwise, it will penetrate the plasma (a). If, on the other hand, the particle is ejected as a low energy ion, it may simply return to the wall (b). This mechanism of impurity removal would be operative in Tokamaks with and without divertors. If the ion has high kinetic energy, and small ejection angle, it will penetrate to the edge of the scrape-off region and then be swept into the divertor (c). From Fig. 21, it may be seen qualitatively, that the influx of impurities could be strongly affected by the charge state, kinetic energy, and angle of emission of the sputtered particles.

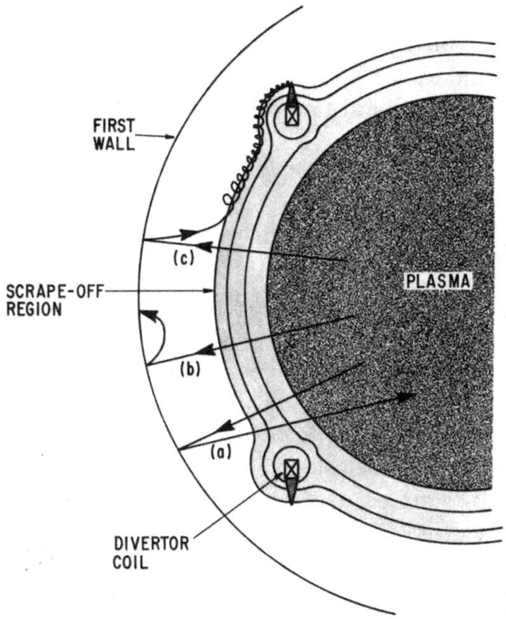

Fig. 21. Schematic cross-section of a divertor-equipped Tokamak showing some of the possible fates of a particle ejected from the wall as: a neutral (*a*), a low energy ion (*b*), and a high energy ion (*c*)[171]

A variety of techniques leading to increased ion fractions are under investigation with a view to assessing the potential of this approach for impurity control in Tokamaks[167–171].

Since oxygen forms an ionic bond with most metals, a metal atom associated with such a bond frequently escapes as an ion when ejected from the surface. Several authors have suggested that this effect be exploited for impurity control in Tokamaks[30, 167–173].

The alkali metals and possibly the alkaline earths are the only elements to form an ionic bond with hydrogen[174], and one might therefore expect that the inevitable presence of surface hydrogen in a reactor environment would further increase the already high ion fraction of these elements. However, not only the ionicity but also the bond strength determine the ion fraction of the sputtered products. Thus for titanium, the positive ion fraction for an oxygen covered surface lies between ~ 20 and 100%[170, 175, 176] and for hydrogen coverage may go as high as ~ 40%. On the other hand for beryllium which may have an ionic molecular bond with hydrogen but is a very weak hydride former, hydrogen has no effect on the secondary ion yield[168].

Still another approach to increasing secondary ion fractions makes use of a technique analogous to high energy discharge cleaning to alter the chemical composition of the near surface region, to a depth of several tens of microns. Such in-situ discharge-induced ion implantation techniques could be used to modify Tokamak surfaces in such a way as to lead to increased ion fraction yields in sputtering.

In the remaining part of this chapter one such approach, the nitriding of metal surfaces, will be discussed in some detail since it could lead both to higher ion fractions and to a lower effective-Z than that of the pure metal.

The nitriding can be easily performed using a low pressure discharge either in pure nitrogen or in a mixture of nitrogen with other gases, generally with hydrogen. The sample to be nitrided can either form the cathode of a discharge ("ion nitriding") or it can be simply immersed in the plasma at a floating potential.

6.5.1 Ion Nitriding

The ion nitriding of steel was investigated fifty five years ago[177)]. Today it represents a well established industrial technology[178–180)]. The nitriding is conventionally performed in a mixture of nitrogen and hydrogen at a total pressure of several mbar. The sample to be nitrided represents the cathode of a dc glow discharge. In spite of numerous studies which were performed, the mechanism of ion nitriding is still not known[178–180)]. Hydrogen has evidently a beneficial effect on the chemisorption of nitrogen atoms on the surface. The nitrogen bulk diffusion appears to be the rate determining step during formation of thick nitride layers. Also the sputtering of the surface followed by redeposition of the sputtered material plays an important role. Nitrided surfaces of titanium and zirconium have been prepared by operating a dc glow discharge in a mixture of 20% N_2 and 80% H_2 at 20–50 mbar[181)]. The discharge was confined between a stainless steel anode and the sample at about 500 mA with a dc potential of 360–400 V. The temperature of the sample was about 900 °C. The kinetics of metal nitride formation are enhanced by about a factor of three in the dc discharge technique compared with either rf discharge or thermal nitriding methods.

The golden-yellow nitrided surfaces were characterized by energy-dispersive x-ray and electron diffraction techniques. The results show that the nitrided surfaces contain mononitride phases. Depth-profile measurements were performed by argon-ion etching and surface analysis by means of x-ray photoelectron spectroscopy.

The kinetics of the formation of surface nitride layers are discussed in terms of a reaction involving nitrogen-hydrogen molecular ions with the substrate. Diffusion of nitrogen into the bulk determines the rate of growth of the nitride layer.

6.5.2 Plasma Nitriding

In earlier work no reaction was observed to occur when metallic samples at floating potential were exposed to nitrogen discharge. It is difficult to elucidate the details of the plasma conditions used in work[180)] on plasma nitriding reported in the literature but two effects appear to be the probable cause of the contradictory results that have been obtained: 1) impurities and 2) a low degree of nitrogen dissociation.

Aubry and Streiff have shown that the rate of nitriding with molecular nitrogen strongly depends on the purity of the nitrogen[182)]. Under ultraclean vacuum conditions nitrogen is dissociatively chemisorbed on metallic surfaces such as Mo, W, and others[183)] and at elevated temperatures it diffuses into the metal[183, 184)]. The saturation coverage of polycrystalline molybdenum and tungsten surfaces is $3–6 \times 10^{18}$ N atoms m^{-2} but the sticking probability decreases by several orders of magnitude before this value is reached[183–185)]. On such a saturated surface there are additional endothermic sites which can be occupied by atomic nitrogen up to a coverage of $\gtrsim 1.2 \times 10^{19} m^{-2}$ [186].

The effect of an intense low pressure discharge in nitrogen is threefold:

1) If the degree of dissociation is high enough the metal surface is readily saturated due to direct adsorption of atoms which can also occupy single isolated sites (the dissociative chemisorption of molecular nitrogen requires two adjacent empty surface sites).

2) Since the sites which are endothermic with respect to N_2 can be populated by N atoms, the total coverage can be significantly higher in the discharge than in molecular nitrogen.

3) The sample at floating potential is charged negatively with respect to the surrounding plasma and the surface is bombarded by ions with energy of a few eV. This, as well as further exothermic processes such as recombination and de-excitation help to keep the surface free of impurities.

More recent experimental work has indeed confirmed that an intense low pressure discharge in nitrogen is an efficient nitriding agent. For example, δ-MoN has been prepared by nitriding of Mo metal for several hours in a nitrogen discharge at a pressure of ~1.8 mbar and a temperature of ~800 °C[187]. The stability of this phase is marginal in 1 atm of molecular nitrogen at this temperature. No data on its stability are known at lower pressures but according to the Le Châtelier-Braun principle, as well as according to the results of Reiter[184] and others[183] the temperature of decomposition for $2\ MoN \rightarrow Mo_2N + 1/2\ N_2$ has to decrease with decreasing nitrogen pressure. Without using a plasma the δ-MoN modification can be prepared only by a long term nitriding of the metal in 1 atm ammonia at ~700 °C[188].

A detailed kinetic study of the nitriding of Mo, Ti and Nb in a nitrogen discharge has been performed recently by Wirz[189] and in Ref.[194, 210]; nitriding of titanium and zirconium was also investigated by Konuma and Matsumoto[190, 191].

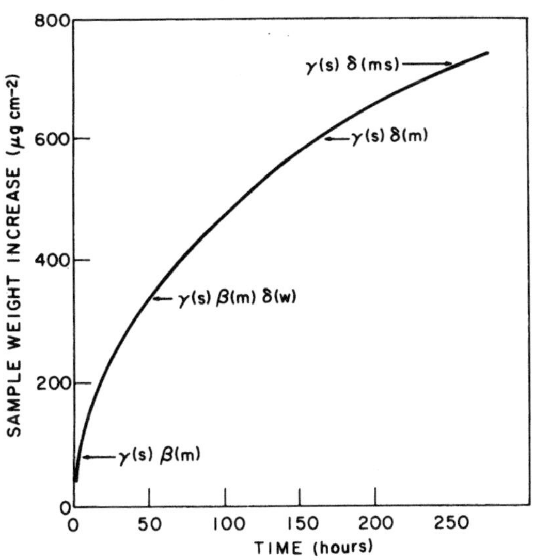

Fig. 22. Appearances of various phases of molybdenum nitride during nitriding (s-strong, m-medium, w-weak). Nitrogen discharge, p = 1.33 mbar, sample temperature 645 °C[189]

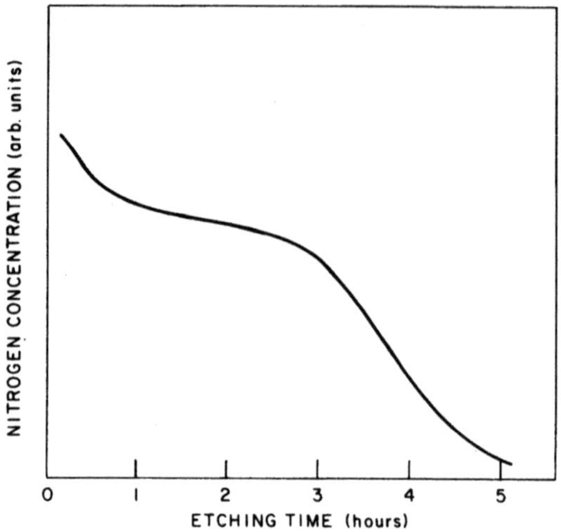

Fig. 23. Depth profile of nitrogen in molybdenum after 91 h of nitriding at 645 °C under the same conditions as given in Fig. 22. During the measurement the surface was sputter etched by Ar^+ bombardment and the composition continuously monitored by Auger spectroscopy[194]

All authors have found that the reaction rates are approximately proportional to the square root of time indicating that bulk diffusion is rate determining. However, the detailed measurements by Wirz have shown that the rate can actually be described by a single rate constant only within a limited time interval. This is mainly due to the complicated depth profile of the nitrogen concentration arising from the stepwise appearance of the various nitride modifications during the nitriding. This is illustrated for molybdenum in Fig. 22 which shows a typical dependence of the sample weight increase on time together with the nitride modifications as found by x-ray diffraction (for details see Ref.[189]). Figure 23 shows the nitrogen depth profile measured on a sample which was nitrided under the same conditions for ~91 h. One sees that there is a thin surface layer of δ-MoN followed by the β-$Mo_2N_{1\pm x}$ and γ-$Mo_2N_{1\pm x}$ modifications. Nitriding of 5 μm thin metal foils has indicated that the reaction rate changes with appearance of the various phases[189].

Similar results were obtained also by nitriding titanium and niobium[210]. Table 5 summarizes the experimental overall rate constants for long term nitriding of thin metal sheets (time of nitriding several hours or more).

Table 5. Experimental kinetic constants for nitriding of titanium, niobium and molybdenum immersed in a low pressure nitrogen discharge at floating potential (according to Ref.[189]). A is the preexponential factor, E_{act} is the activation energy

	A [$\mu g^2 m^{-4} s^{-1}$]	E_{act} [kcal $mole^{-1}$]
Ti	1.6×10^{18}	44.2
Nb	2.9×10^{12}	24.3
Mo	1.45×10^{15}	31.0

As pointed out above, stainless steel is of primary interest for the near term large Tokamaks. Little work has been done on the nitriding of stainless steel by this method so far. Measurements performed by the Zurich group[192)] revealed that the reaction rates are an order of magnitude higher than those for titanium at ~600 °C (see Table 5). However, nitrided steel is unstable under exposure to hydrogen plasma[194a)].

In order to be of interest as protective coatings for Tokamaks the nitride layers have to possess reasonable stability upon bombardment with energetic ions, against thermal shocks, and be permeable to hydrogen in order to avoid a buildup of hydrogen in the bulk material which would cause flaking. By careful analysis of the x-ray diffraction pattern Wirz[189)] has shown that a large stress buildup occurs in the surface during nitriding. This may negatively influence the stability of such a layer but no consistent experimental data are available yet. Veprek, et al.[207)] have observed flaking of the nitride layer upon bombardment with MeV He^+ at a dose of $\sim 2 \times 10^{22}$ ions m^{-2}, which is in the range of the critical dose for blistering of metals[41)].

The nitride layer on titanium and niobium reduces the rate of diffusion of atomic hydrogen into the underlying metal. The measurements were performed by following the weight changes of samples immersed in a hydrogen low pressure discharge and by x-ray diffraction from the sample surface[193,194)]. Depending on the experimental conditions, a reduction of the hydrogen diffusion rate up to two orders of magnitude has been observed for the nitrided sample as compared with the untreated metal. The rate of hydrogen diffusion through the nitride layer obviously depends on a number of parameters, such as the crystallinity of the metal, the conditions of nitriding, the thickness of the nitride layer and also on the temperature during nitriding, but more detailed studies are necessary in order to obtain consistent data.

Similar measurements were also performed on stainless steel. The nitride layer was attacked by the hydrogen discharge and therefore no quantitative data on the hydrogen diffusion could be obtained[192, 194a)]. This system also requires further investigation.

7 Conclusions

Plasma impurities are recognized to be a limiting factor in the performance of present day Tokamaks. Unless solutions to the impurity problem can be found, further progress towards fusion as an energy source will be seriously jeopardized.

More data are needed on erosion rates by physical and physichemical sputtering, hydrogen trapping, reflection, retention and desorption, on structural changes due to radiation damage and to long term exposure to hydrogen plasma. In addition, new materials have to be developed and characterized with regard to their performance in fusion plasma environments.

The control of impurity release and transport requires a better understanding of the complex phenomena of plasma-wall interactions including the processes occuring in the scrape-off layer in the limiter shadow. In order to establish the feasibility of suggested solutions such as divertors or surface modifications, experiments have to be performed not only in the laboratory but also in-situ in fusion devices. The latter includes the study of trapping efficiencies, recycling, secondary ion and neutral emission and various desorption phenomena.

Most of the studies performed so far have yielded data on individual processes and there is almost no information on synergistic effects. For example, measurements of sputtering yields have to be performed under conditions where the surface is simultaneously exposed to low energy and thermal particles, uv- and x-ray radiation, etc.

Discharge cleaning, although routinely used to obtain plasmas almost free of low-Z impurities, is not yet sufficiently understood and fundamental investigations could help to improve its efficiency.

Protective coatings of limiters, beam dumps and first walls can significantly relax the material requirements with respect to plasma-wall interactions and structural integrity. Although coatings for near term experimental devices may be prepared using state of the art technology, coatings in a future power reactor will probably require the development of new techniques, preferably permitting in-situ deposition.

The importance of the impurity problems discussed in this chapter are fully recognized as evidenced by the fact that several Tokamaks are under construction or in operation with significant machine time devoted to study impurity release, transport and controls.

8 References

1. Fusion Power by Magnetic Confinement, Program Plan. U. S. Energy Research & Development Administration ERDA-76/110. Report available from Nat. Tech. Information Service (NTIS), U. S., Dep. of Commerce, Springfield, Virgina 22161
2. Fusion Reactor Design II. Progress Summary, Conclusions and Prognosis on Fusion Reactors: Second IAEA Technical Committee Meeting and Workshop; Conn, R. W., Frank, T. G., Hancox, R., et al., Madison: Univ. of Wisconsin October 1977. Vienna: Int. Atomic Energy Agency 1978
3. Steiner, D.: Nuclear Sci. and Eng. *58*, 107 (1975)
4. Ribe, F. L.: Rev. Mod. Phys. *47*, 7 (1975)
5. Chemistry of Fusion Technology, Gruen, D. M. (ed.), New York: Plenum Press 1972
6. Conn, R. W., et al.: Proc. 6th IAEA Conf. on Plasma Phys. and Control. Nucl. Fusion Res., Berchtesgaden, October 1976
7. Usher, J. L., Powel, J. R., Fillo, J. A., Lazareth, O. W.: In: Proc. 9th Symp. on Fusion Technology. (Garmisch-Partenkirchen 1976). p. 387. Publ. by Commission of the European Communities, Oxford: Pergamon Press 1976
8. Lawson, D. J.: Proc. Phys. Soc. London *B 70*, 6 (1957)
9. Eubank, H. P.: In: 20th Annual Meeting of the Division of Plasma Physics of the American Physical Society (Colorado Springs, Colorado 1978)
10. World Survey of Major Facilities in Controlled Fusion Research. Vienna: Int. Atomic Energy Agency 1976
11. Moss, J.: Nucl. Eng. Int. *22*, 33 (1977)
12. Bickerton, R. J.: J. Nucl. Materials *76* and *77*, 16 (1978)
13. Simonen, T. C., et al.: J. Nucl. Materials *63*, 59 (1976)
14. Dimov, G. I., Zakaidov, V. V., Kishnivsky, M. E.: In: Proc. 6th Int. Conf. on Plasma Phys. and Control Nucl. Fusion, (Berchtesgaden 1976). p. 177. Vienna: IAEA, 1977
15. Post, R. F.: J. Nucl. Materials *76* and *77*, 112 (1978)

16. Jassby, D. L.: Nucl. Fusion *17,* 2, 396 (1977)
17. Lehner, G.: In: Reactions under Plasma Conditions, Vol. II, Venugopalan, M. (ed.), New York: Wiley Interscience 1971
18. Ribe, F. L., Krakowski, R. A., Thomassen, K. I., Coultas, T. A.: Nucl. Fusion, Special Suppl. on Fusion Res. Design Problems, 1974
19. A Joint Report by Argonne Nat. Lab. and Los Alamos Sci. Lab. on Eng. Design Study of a Reference Theta-Pinch Reactor 1977–82
20. Engineering Development Program Plan for Inertial Confinement Fusion, Volume II, Program Strategy, PNL-2582. Battelle, Pacific Northwest Laboratories, Richland, Washington, U. S., 1978
21. Sandia Technology, SAND76-0615, Vol. 2, No. 3. Sandia Laboratories, Albuquerque, New Mexico, U. S., 1976
22. Artsimovich, I. V.: Nucl. Fusion *12,* 215 (1972)
23. Cohen, S. A.: J. Vac. Sci. Technol. *13,* 449 (1976)
24. Conn, R. W.: Nucl. Eng. Dep., Univ. of Wisconsin, Report UWFDM-175 (1976). Published In: Proc. 2nd ANS Topical Meeting on the Technology of Control Nucl. Fusion. Richland, WA: September, 1976
25. Badger, B., Conn, R. W., Kulcinski, G. L., et al.: UWMAK-I, A Wisconsin Toroidal Fusion Reactor Design. Nucl. Eng. Dept. Univ. of Wisconsin, Report UWFDM-68. Wisconsin, 1973. See also: Kulcinski, G. L. and Conn, R. W. in: Fusion Reactor Design Problems. Vienna: IAEA, 1974
26. Badger, B., Conn, R. W., Kulcinski, G. L., et al.: UWMAK-II, A Conceptual Tokamak Power Reactor Design. Nucl. Eng. Dept., Univ. of Wisconsin, Report UWFDM-112. Wisconsin,1975
27. Badger, B., Conn, R. W., Kulcinski, G. L., et. al.: UWMAK-III, A Noncircular Tokamak Power Reactor Design. Nucl. Eng. Dept., Univ. of Wisconsin, Report UWFDM-150. Wisconsin, 1976
28. Conn, R. W.: J. Nucl. Materials *76* and *77,* 103 (1978)
29. Eckhartt, D. and Venus, E.: JET Technical Note *9* (1974)
30. Behrisch, R.: J. de Physique *38,* C3 (1977)
31. Hopkins, G. R.: Proc. 5th Conf. on Plasma Phys. and Control. Nucl. Fusion (Tokyo 1974), Vol. II., p. 275. Vienna: IAEA, 1975
32. Conn, R. W., Houlberg, W. A., Kesner, J.: Nucl. Eng. Dept., Univ. of Wisconsin, Report UWFDM-106. Wisconsin, 1975
33. Vernickel, H.: Phys. Reports *37,* 93 (1978)
34. Conn, R. W. and Kesner, J.: Nucl. Fusion *15,* 775 (1975)
35. Meade, D. M.: Nucl. Fusion *14,* 289 (1974)
36. Behrisch, R., Kadomtsev, B. B.: Proc. 5th Conf. on Plasma Phys. and Control. Nucl. Fusion (Tokyo 1974) Vol. II, p. 229. Vienna: IAEA, 1975
37. Griem, H. R.: In: Ref.[39)], p. 193
38. Proc. 2nd Int. Conf. Surface Effect in Control. Fusion Devices (San Francisco 1976) Bauer, W., Fingfeld, C. R., Kaminsky, M. (ed.). Amsterdam: North-Holland Publ. Co. 1976. J. Nucl. Materials *63* (1976)
39. Proc. Int. Symp. on Plasma Wall Interaction (Jülich 1976). Oxford: Pergamon Press 1977
40. Proc. 3rd Int. Conf. Surf. Effects in Control. Fusion Devices (Culham 1978). McCracken, G. M., Stott, P. E., Thompson, M. W. (ed.). Amsterdam: North-Holland Publ. Co. 1978; J. Nucl. Materials *76* and *77* (1978)
41. Scherzer, B. M. U., Behrisch, R., Roth, J.: In: Ref.[39)], p. 353
42. Bohdansky, J., Roth, J., Sinha, M. K.: Ref.[7)], p. 541
43. Vepřek, S., Haque, R. M.: Appl. Phys. *8,* 303 (1975)
44. Roth, J., Bohdansky, J., Poschenrieder, W., Sinha, M. K.: J. Nucl. Materials *63,* 222 (1976)
45. Erents, S. K., Braganza, C. M., McCracken, G. M.: J. Nucl. Materials *63,* 399 (1976)
46. Busharov, N. P., Gorbatov, E. A., Gusev, V. M., Guseva, M. I., Martynenko, Yu. V.: J. Nucl. Materials *63,* 230 (1976)
47. Vepřek, S., Webb, A. P., Oswald, H. R., Stüssi, H.: J. Nucl. Materials *68,* 32 (1977)
48. Bohdansky, J., Bay, H. L., Ottenberger, W.: J. Nucl. Materials *76* and *77,* 163 (1978)

49. Rutherford, P. H., Hirshman, S. P., Jensen, R., Post, D. E., Seidl, F. G. P.: In: Ref.[39], p. 173
50. Arunasalam, V., et al.: Proc. 8th European Conf. on Control. Fusion and Plasma Phys., (Prague, 1977) Vol. 1, p. 17
51. Hugill, J., et al.: ibid, Vol. 1, p. 39 and Vol. 2, p. 49
52. Colchin, R. J., et al.: J. Nucl. Materials *76* and *77*, 405 (1978)
53. Draley, J. E., Frost, B. R. T., Gruen, D. M., Kaminsky, M., Maroni, V. A.: Proc. 1971 Intersociety Energy Conversion Eng. Conf. p. 1065. New York: Soc. of Automotive Eng. 1971
54. Taylor, R. J.: J. Nucl. Materials *76* and *77*, 412 (1978)
55. Report by the T. F. R.-group: In: Ref.[39], p. 3
56. Report by the Petulla-group: In: Ref.[39], p. 105
57. Ohasa, K., et al.: J. Nucl. Materials *76* and *77*, 489 (1978)
58. Ginot, P.: J. Nucl. Materials *76* and *77*, 30 (1978)
59. Hinnov, E.: J. Nucl. Materials *53*, 16 (1974)
60. Pospieszczyk, A., Burt, J., Fielding, S. J., McCracken, G. M. and Scott, P. E.: In: Ref.[39], p. 471
61. Marmar, E. S.: J. Nucl. Materials *76* and *77*, 59 (1978)
62. Behrisch, R.: In: Critical Materials Problems in Energy Production, Stein, C. (ed.), p. 91. New York: Academic Press 1976
63. Weismann, R., Sigmund, P.: Rad. Effects *9*, 7 (1973)
64. Bøttiger, J., Winterbon, K. B.: Rad. Effects *20*, 65 (1973)
65. Oen, O. S., Robinson, M. T.: Nucl. Instr. and Meth. *132*, 647 (1976)
66. Robinson, J. E.: Rad. Effects *23*, 29 (1974); Proc. 3rd Inter. Conf. Atomic Col. with Solids, Kiev, 1974
67. Ishitani, T., Shimizu, R., Murata, K.: Jap. J. Appl. Phys. *11*, 125 (1972)
68. Barnett, C. F., Ray, J. A.: Nucl. Fusion *12*, 65 (1972)
69. Matschke, F. E. P., Eckstein, W., Verbeek, H.: Verh. DPG IV *10*, 47 (1975)
70. Eckstein, W., Matschke, F. E. P., Verbeek, H.: J. Nucl. Materials *63*, 199 (1976)
71. Sidenius, G.: Phys. Lett. *49-A*, 409 (1974); and Nucl. Instr. and Meth. *132*, 673 (1976)
72. Bohdansky, J., Roth, J., Sinha, M. K., Ottenberger, W.: J. Nucl. Materials *63*, 115 (1976)
73. Lindhard, J., Scharff, M., Schiott, H.: Mat. Fys. Medd. *33*, No. 14, (1963)
74. Verbeek, H.: J. Appl. Phys. *46*, 2981 (1975)
75. Gruen, D. M., Varma, R., Wright, R. B.: J. Chem. Phys. *64*, 5000 (1976)
76. Picraux, S. T., Bøttiger, J., Rud, N.: J. Nucl. Materials *63*, 110 (1976)
77. Bottiger, J., Picraux, S. T., Rud, N., Lausen, T.: J. Appl. Phys. *48*, 926 (1977)
78. Oriani, R. A.: Acta Metall. *18*, 147 (1970)
79. Bottiger, J., Leslie, J. R., Rud, N.: J. Appl. Phys. *47*, 1672 (1976)
80. Bauer, W., J. Nucl. Materials, *76* and *77*, 3 (1978)
81. Cohen, S. A., Dylla, H. F., Rossnagel, S. M., Picraux, S. T., Borders, J. A., Magee, C. W.: J. Nucl. Materials *76* and *77*, 459 (1978)
82. Altstetter, C., Behrisch, R., Bottiger, J., Pohl, F., Scherzer, B. M. U.: Nucl. Instr. and Methods *149*, 59 (1978)
83. Wilson, K. L., Baskes, M. I.: J. Nucl. Materials *76* and *77*, 291 (1978)
84. Langley, R. A., Blewer, R. S., Roth, J.: J. Nucl. Materials *76* and *77*, 313 (1978)
85. Erents, S. K.: In: Proc. Int. Conf. Applications of Ion Beams to Materials, Univ. Warwick 1975
86. Cohen, S. A.: Private communication
87. Hotston, E. S., McCracken, G. M.: J. Nucl. Materials *68*, 277 (1977)
88. Borders, J. A., Langley, R. A., Wilson, K. L.: J. Nucl. Materials *76* and *77*, 168 (1978)
89. Smith, D. L.: J. Nucl. Materials *75*, 20 (1978)
90. Rausch, E. O., Bazhin, A. I., Thomas, E. W.: J. Chem. Phys. *65*, 4447 (1976)
91. Roth, J. Bohdansky, J., Poschenrieder, W., Sinha, M. K.: J. Nucl. Materials *63*, 222 (1976)
92. Busharov, N. P., et al.: Report I. V. Kuchatov, Inst. Atomic Energy (Moscow, 1975). Engl. transl. ERDA-Tr-50

93. Braganza, C., McCracken, G. M., Erents, S. K.: Proc. Int'l. School on Physics of Ionized Gases, August 27, 1976, Dubrovnik, Yougoslavia. See also J. Nucl. Materials *76* and *77*, 204 (1978)
94. Smith, J. N., et al.: J. Nucl. Materials *63*, 393 (1976)
95. Guntherschulze, A.: Vacuum *3*, 360 (1953)
96. Veprek, S.: J. Crystal Growth *17*, 101 (1972)
97. Balooch, M., Olander, D. R.: J. Chem. Phys. *63*, 4772 (1975)
98. Gruen, D. M., Wright, R. B., McBeth, R. L., Sheft, I.: J. Chem. Phys. *62*, 1192 (1975)
99. McAllister, J. W., White, J.: J. Chem. Phys. *58*, 1496 (1973)
100. Lichtman, D.: J. Nucl. Materials *53*, 285 (1974)
101. Drinkwine, M. J., Shapira, Y., Lichtman, D.: In: Adv. in Chemistry Series *158*, Kaminsky, M. (ed.), p. 171. Washington D. C.: Amer. Chem. Soc. 1976
102. Lichtman, D., Shapira, Y.: J. Nucl. Materials *63*, 184 (1976)
103. Brumbach, S., Kaminsky, M.: J. Nucl. Materials *63*, 188 (1976)
104. Brumbach, S., Kaminsky, M.: J. Appl. Phys. *47*, 2844 (1976)
105. Winters, H. F., Sigmund, P.: J. Appl. Phys. *45*, 4760 (1974)
106. Taglauer, E., Beital, V., Marin, G., Heiland, W.: J. Nucl. Materials *63*, 193 (1976)
107. McCracken, G. M.: Vacuum *24*, 463 (1974)
108. Das, S. K., Kaminsky, M.: In: Ref.[101)], p. 112
109. Kaminsky, M. S., Das, S. K.: Rad. Effects *18*, 245 (1973)
110. Erents, S. K., McCracken, G. M.: Rad. Effects *18*, 191 (1973)
111. Bauer, W., Thomas, G. J.: J. Nucl. Materials *53*, 127 (1974)
112. Wilson, K. L., Haggmark, L. G., Langley, R. A.: In: Ref.[39)], p. 401
113. Roth, J., Behrisch, R., Scherzer, B. M. U.: J. Nucl. Materials *57*, 365 (1975)
114. McCracken, G. M., Goodall, D. H. J.: UKAEA preprint CLM-P-509, Culham Laboratory, October 1977
115. Ali-Kahn, A., Dietz, K. J., Waelbroeck, F. G., Wienhold, P.: J. Nucl. Materials *76* and *77*, 337 (1978)
116. Cafasso, F., Gruen, D., Kaminsky, M., Robinson, J. E., Wiedersich, H.: Nuclear Technology *34*, 131 (1977)
117. Spitzer, L.: Phys. Fluids *1*, 253 (1950)
118. Maede, D. M.: In: Ref.[39)], p. 21
119. Stott, P. E., Burt, J., Erents, S. K., et al.: Ref.[39)], p. 39
120. Shimomura, Y., Maeda, H.: J. Nucl. Materials *76* and *77*, 48 (1978)
121. Mathewson, A. G.: In: Ref.[39)], p. 517
122. Grieger, G., et al.: Proc. 6th Europ. Conf. on Control. Fusion and Plasma Phys., Vol. 1, p. 101. Moscow, 1973
123. W VII A Team, Blaumoser, M., et al.: Proc. 6th Conf. on Plasma Phys. and Control. Nucl. Fusion (Berchtesgaden 1976) and Plasma Physics and Control. Nucl. Fusion Research, Vol. II, p. 81. IAEA, Vienna 1977
124. Taylor, R. J.: J. Nucl. Materials *76* and *77*, 41 (1978)
125. Dietz, K. J., Waelbroeck, F. G.: In: Ref.[39)], p. 445
126. Dietz, K. J., Waelbroek, F. G., Wienhold, P.: Proc. 3rd Int. Symposium on Plasma Chemistry. Limoges: University of Limoges 1977, paper No. G.2.5
127. TFR-Group: J. Nucl. Materials *76* and *77*, 587 (1978)
128. v. Seggern, J., Tschersich, K. G.: J. Nucl. Materials *76* and *77*, 600 (1978)
129. Holland, L., Rizk, A. S.: J. Nucl. Materials *76* and *77*, 605 (1978)
130. Stacey, W. M., et al.: Tokamak Experimental Power Reactor Studies. CTR Program, Argonne Nat. Lab., Report ANL/CTR-75-2. Argonne 1975
131. Stacey, W. M., et al.: ibid, Report ANL/CTR-76-3. Argonne 1976
132. Stacey, W. M., Smith, D. L., Brooks, J. N.: In: Ref.[39)], p. 599
133. Smith, D. L.: In: First Wall Coating Workshop. Mattox, D. M. (ed.). Albuquerque: Sandia Laboratory 1978
134. Vepřek, S., Haque, M. R., Oswald, H. R.: J. Nucl. Materials *63*, 405 (1976)
135. Chin, J., Ohkawa, T.: Nucl. Technology *32*, 115 (1977)

136. First Wall Coating Workshop. Mattox, D. M. (ed.). Albuquerque: Sandia Laboratory 1978
137. Yee, K. K.: Int. Metal Rev. *226*, 19 (1978)
138. Tarassenko, S.: The proposed bake-out system for the JET vacuum vessel. Report of the JET Group JTN/C (FF) 105
139. Gerdeman, D. A., Hecht, N. L.: Arc plasma technology in material science. Wien: Springer 1972
140. Fauchais, P., Bourdin, E.: This book, Vol. III
141. Kingcade, J. E., et al.: In: Ref.[39], p. 263
142. Braganza, C., Vepřek, S., Groner, P.: J. Nucl. Materials *85* and *86*, 1133 (1979)
143. Rovner, L. H., Chen, K. Y., Chin, J.: Proc. ANL Topical Meeting on the Technology of Controlled Nuclear Fusion. Santa Fe 1978
144. Techniques and Applications of Plasma Chemistry. Hollahan, J. R. and Bell, A. T. (ed.). New York: Wiley 1974
145. Küppers, D., Koenings, J., Wilson, H.: J. Electrochem. Soc. *123*, 1079 (1976)
146. Spear, W. E., LeComber, P.: In: Amorphous and Liquid Semiconductors. Spear, W. E. (ed.), p. 309. Edinburgh: University of Edinburgh 1977
147. Vepřek, S.: Pure Appl. Chem. *48*, 163 (1976)
148. Braganza, C., McCracken, G. M., Erents, S. K.: In: Ref.[39], p. 257
149. Anderson, D. A., Spear, W. E.: Phil. Mag. *35*, 1 (1977)
150. Bunshah, R. F.: In: Proc. 7th Int. Vac. Congress and 3rd Int. Conf. Solid Surf. Dobrozemsky, R., et al. (eds.), p. 1553. Vienna 1977
151. Kordis, J., Vepřek, S.: Report No. VII/1977. Zürich: Inst. Inorg. Chem., University of Zürich 1977
152. McCracken, G. M., Maple, J. H. C.: Brit. J. Appl. Phys. *18*, 919 (1967)
153. McCracken, G. M., Jefferies, D. K., P. Goldsmith, P.: Proc. Fourth int. Vac. Congr., Part I., 149 (1968)
154. McCracken, G. M., Jefferies, D. K.: 5th Symp. on Fusion Techn., Oxford, Paper No. 149 (1968)
155. McCracken, G. M., Erents, S. K.: B. N. E. S. Nuclear Fusion Reactor Conf., Culham, England, Paper No. 4.2, p. 353 (1969)
156. Sheft, I., Reis, A., Gruen, D. M., Peterson, S. W.: J. Nucl. Materials *59*, 1 (1976)
157. Yonts, O. C., Strehlow, R. A.: J. Appl. Phys. *33*, 2903 (1962)
158. Schiøtt, H. E.: I. Dansk. Vidensk. Selsk. Mat.-Fys. Medd. *35*, No. 9 (1968)
159. Sawatsky, A.: J. Nucl. Materials *9*, 363 (1963)
160. Zamir, D., Cotts, R. M.: Phys. Rev. *134*, A666 (1964)
161. Marshall, R. P.: Trans. Met. Soc., A.I.M.E. *233*, 1449 (1965)
162. Albrecht, W. M., Goode, W. D., Mallett, M. W.: Battelle Memorial Inst. Rep., No. B.M.I. 1332 (1959)
163. Morton, J. R., Stark, D. S.: Trans. Faraday Soc. *56*, 354 (1960)
164. Mueller, W. M., Blackledge, J. P., Libowitz, G. G. (eds.): Metal Hydrides, New York: Academic Press 1968
165. Bau, R. (ed.): Transition Metal Hydrides, Advances in Chemistry Series, No. 167, American Chemical Society, Washington, D. C., 1978
166. Behrisch, R., Kadomtsev, B. B.: Nucl. Fusion Spec. Suppl. 451 (1974)
167. Krauss, A. R., Gruen, D. M.: J. Nucl. Materials *63*, 380 (1976)
168. Krauss, A. R., Gruen, D. M.: Surf. Sci. *90*, 564 (1979)
169. Krauss, A. R., Gruen, D. M.: Appl. Phys. *14*, 89 (1977)
170. Krauss, A. R., Gruen, D. M.: Nucl. Inst. Meth. *149*, 547 (1978)
171. Krauss, A. R., Gruen, D. M.: J. Nucl. Materials *85* and *86*, 1179 (1979)
172. Hofer, W. O., Bay, H. L., Martin, P. J.: J. Nucl. Materials *76* and *77*, 156 (1978)
173. Smith, Jr., J. N.: J. Nucl. Materials *80*, 356 (1979)
174. Mackay, K. M.: Hydrogen Compounds of the Metallic Elements, pp. 20–21. (SPON, 1966)
175. Benninghoven, A.: Surf. Sci. *53*, 596 (1975)
176. Steinbruchel, C., Gruen, D. M.: J. Vac. Sci. Technol. *16*, 251 (1979)
177. Fry, A.: Stahl Eisen *43*, 1279 (1923): J. Iron Steel Inst. *125*, 197 (1932)

178. Hudis, M.: In: Techniques and Applications of Plasma Chemistry. Hollahan, J. R. and Bell, A. T. (eds.). New York: Wiley 1974
179. Seybolt, A. U.: Trans. Metallurg. Soc. AIME *245*, 769 (1969)
180. Hudis, M.: J. Appl. Phys. *44*, 1489 (1973)
181. Liu, M. B., Gruen, D. M., Krauss, A. R., Reis, Jr., A. H., Peterson, S. W.: High Temp. Sci. *10*, 53 (1978)
182. Aubry, J., Streiff, R.: J. Electrochem. Soc. *118*, 650 (1971)
183. Interactions on Metal Surfaces, Gomer, R. (ed.). Berlin: Springer 1975
184. Reiter, F.: In: Ref.[7]), p. 57
185. Huang, Ch.: Chemisorption of Hydrogen and Nitrogen on the Mo (100) Surface. Ph. D. Thesis. Providence, R. I.: Brown University 1974
186. Winters, H. F., Horne, D. E.: Surf. Sci. *24*, 587 (1971)
187. Vepřek, S.: (unpublished results 1970)
188. Ettmayer, P.: Monatshefte f. Chemie *101*, 127 (1970)
189. Wirz, E.: Ph. D. Thesis. Zürich: University of Zürich 1979
190. Konuma, M., Matsumoto, O.: J. Less-Common Metals *52*, 145 (1977)
191. Konuma, M., Matsumoto, O.: ibid *55*, 97 (1977)
192. Braganza, C., Vepřek, S.: (unpublished results 1978)
193. Braganza, C., Cordis, J., Vepřek, S.: J. Nucl. Materials *76* and *77*, 612 (1978)
194a. Braganza, C., Stüssi, H., Vepřek, S.: J. Nucl. Materials *87*, 331 (1979)
194b. Braganza, C., Vepřek, S., Wirz, E., Stüssi, H., Textor, M.: Proc. 4th Int. Symp. on Plasma Chemistry (Zürich, 1979), Vepřek, S. and Hertz, J. (ed.), Vol. 1, p. 100, University of Zürich 1979
195. Behrisch, R., Blewer, R. S., Kurkal, H., et al.: J. Nucl. Materials *76* and *77*, 437 (1978)
196. Webb, A. P., Vepřek, S.: Chem. Phys. Letters *62*, 173 (1979)
197. Vepřek, S., Webb, A. P., Stüssi, H.: In: Ref.[39]), p. 431
198. Vepřek, S., Portmann, A., Webb, A. P., Stüssi, H.: Radiation Effects *34*, 183 (1977)
199. Vepřek, S., Webb, A. P., Portmann, A., Stüssi, H.: Proc. 3rd Int. Conf. on Solid Surf., p. 1505, Vienna 1977
200. Stüssi, H., Webb, A. P., Vepřek, S.: J. Nucl. Materials *76* and *77*, 636 (1978)
201. Brunner, J., Thuler, M., Stüssi, H., Vepřek, S., Webb, A. P.: Rad. Effects *38*, 191 (1978)
202. Stüssi, H., Vepřek, S., Webb, A. P.: Rad. Effects Letters *43*, 133 (1979)
203. Rao, A. S., Bacon, D. J.: Proc. 5th London Int. Carbon and Graphite Conf., London 1978
204. Das, S. K., Kaminsky, M., Tishler, R. and Cecchi, J.: J. Nucl. Materials *85* and *86*, 225 (1979)
205. Roth, J. (private communication)
206. Vepřek, S., Webb, A. P., Oswald, H. R., Stuessi, H.: J. Nucl. Materials *68*, 32 (1977)
207. Vepřek, S., Stuessi, H., Braganza, C.: 1978 (unpublished)
208. McCracken, G. M., Stott, P. E.: Nucl. Fusion *19*, 889 (1979)
209. Proc. 4th Int. Conf. Surf. Effects in Control. Fusion Devices (Garmisch-Partenkirchen 1980) (to be published)
210. Wirz, E., Oswald, H. R., Vepřek, S.: Proc. 4th Int. Symp. on Plasma Chemistry (Zürich 1979), Vepřek, S., Hertz, J. (ed.), Vol. 2, p. 492, University of Zürich 1979
211. Wieder, H., Cardona, M., Garnieri, C. R.: Phys. Stat. Sol. *92*, 99 (1979)

Received April 20, 1979

Preparation of Optical Waveguides
with the Aid of Plasma-Activated Chemical Vapour Deposition at Low Pressures

D. Küppers and H. Lydtin

Philips GmbH Forschungslaboratorium Aachen, D-5100 Aachen, F.R.G.

Table of Contents

1 Introduction

Until recently optical communications were restricted by the lack of fast monochromatic light sources and sensitive photodetectors. Prospects for optical communications improved considerably about two decades ago when a powerful light source became available with the invention of the laser. After that, the transmission medium was the bottleneck of an optical communication system. At that time an intensive search for a new transmission medium was started, particulary because free space propagation could be ruled out for civil use as a consequence of the relative frequent occurrence of atmospheric disturbances.

At a fairly late stage in 1966 optical fibers were first considered seriously to be a possible transmission medium by Kao and Hockham[1)]. In 1969 Jones and Kao[2)] demonstrated that minimum attenuation losses in bulk silica can be as low as 5 db/km. The actual breakthrough was achieved in 1970 with the preparation of fibers with optical losses below 20 db/km[3)]. Since then rapid progress has been made in this field all over the world.

Today fiber-optic transmission systems offer several advantages over conventional copper wire and coaxial cable systems. Among these are increased bandwidth, smaller size, lower weight, lack of crosstalk, and a very low susceptibility to electromagnetic interference. It is to be expected that these advantages will open widespread application of fiber-optic transmission systems in the future. This seems to be supported by the fact that a great number of public and in-house trial systems are under test all over the world.

2 Optical Fibers for Telecommunication

Figure 1 shows a schematic view of an optical telecommunication line. Such a line consists of a source, normally a GaAs laser or a light-emitting diode, typically operating in the wavelength region between 0.8–0.9 μm, a transmission line, the optical fiber, and a receiver, typically an avalanche photodiode. As can be seen from Fig. 1, the fiber guides the light in a central region, called the core. This core has a refractive index which is somewhat higher than the surrounding cladding. Because of the greater refractive index the light is confined to this region. For fibers with losses near the intrinsic minimum of attenuation, it is important that a cladding protects the light from dust particles and other impurities present on the outer surface of the fiber.

Systems as shown in Fig. 1 can be used in several applications which differ in the information that can be transmitted per unit time and unit length of the transmission line. High-quality fibers as obtained, for example, by means of the nonisothermal plas-

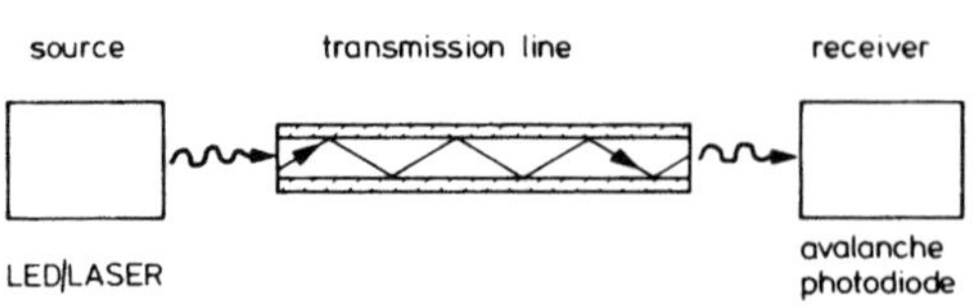

Fig. 1. Schematic optical communication system

ma deposition process are intended for high-capacity and long-distance transmission. To give an impression of the capability of these fibers it should be realized that 15,000 telephone calls can be simultaneously transmitted in one single fiber over a length of 1 km. A conventional coaxial cable (9.5 mm o.d.) is able to carry 10,800 telephone channels at the same time over a distance of 1.55 km. An optical cable, having the same geometric dimensions as a coaxial cable, allows a package of at least 15 fibers and this means that the information carrying capacity will be about 20 times larger.

In the following, the light guiding properties will be discussed that are important for optical fibers and that possibly impose some constraints on the preparation process.

2.1 Attenuation and Pulse Broadening in Optical Fibers

The most important requirements to be met by optical fibers are low optical losses and high transmission bandwidth. Sources of attenuation are absorption and scattering. Intrinsic absorption losses in the ultraviolet region are caused by electronic transitions. In the visible and infrared, Rayleigh scattering and excitation of vibrations are the dominant loss mechanisms. In between the visible and infrared region there is a transmission window which is attractive for optical communications. In this wavelength region absorption is sensitive to trace impurities of the transition elements, as indicated in Fig. 2, and to water incorporation. As can be seen from this Figure, impurity contents of transition elements in the order of some ppb may raise the absorption losses by several db/km.

With respect to attenuation the demands on the preparation methods[5] are just as stringent as in semiconductor fabrication. This means that ultrapure starting substances must be used. Fortunately, it turns out that most of the glass-forming oxides, such as SiO_2, GeO_2, B_2O_3, P_2O_5 etc., can be obtained from commercially available semiconductor-grade liquids and gases.

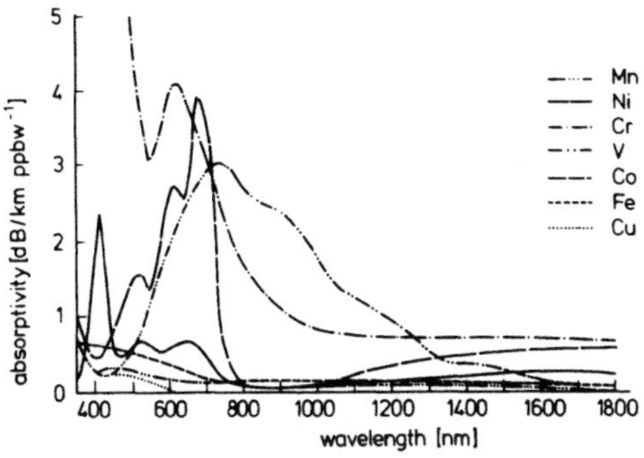

Fig. 2. Practical absorptivity[1] spectra for transition elements in fused silica prepared by flame hydrolysis; absorptivity is given in terms of ppbw metal[4]. (Reproduced by permission of The American Ceramic Society)

1 The practical absorptivities indicate the overall effect of the transition element when it is incorporated into fused silica by the flame hydrolysis method. The absorptivities do not reveal the absorbing power of the individual valence state (i.e. true absorptivity).

The second source of attenuation is the intrinsic scattering of light caused by density fluctuations of the glass on a microscopic scale. This type of scattering loss may be increased by additional inhomogeneities such as small bubbles, striations, imperfections at the core/cladding interface, and others. In a well-prepared fiber all additional losses are small compared with intrinsic losses. Figure 3 shows the attenuation achieved up to now for wavelengths near the transmission window.

For high transmission capacities pulse broadening during the time of flight of an optical signal travelling through the fiber should be as small as possible. The main causes of the broadening are mode dispersion and material dispersion. Mode dispersion depends on the actual path of the light through the fiber and can be influenced by the index profile in the core of the fiber. Three different fiber types are illustrated in Fig. 4. For a step index fiber the maximum delay time between fast on-axis and slow off-axis rays is given by[7)]

$$\tau = N_1 \cdot \Delta \cdot \frac{l}{c}$$

where

$$N_1 = n_1 - \lambda \frac{dn_1}{d\lambda} \qquad (1)$$

is the material group index in the fiber core and where Δ is the relative difference of the refractive index between core and cladding. l is the fiber length and c is the vacuum velocity of light. For a one per cent difference in refractive index between the

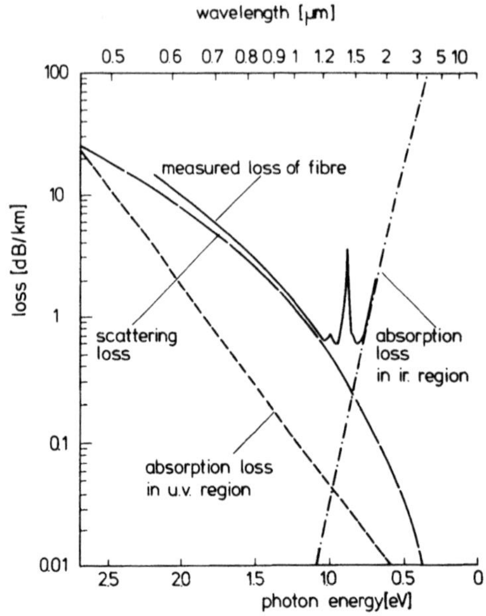

Fig. 3. Spectral loss curve of germania-doped-silica-glass-core fiber separated into inherent u. v. absorption loss[2], scattering loss and inherent i.r. loss[6)]. (Reproduced by permission of IEE)

2 Instead of "absorption loss in u.v. region" the lettering in the Fig. 3 should read "absorption loss as extrapolated from the u.v. region" as is obvious from the text[6)].

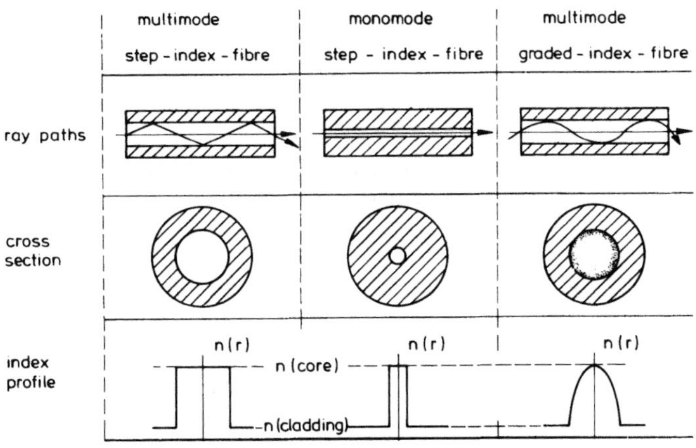

Fig. 4. Ray paths, cross section and refractive index profiles in different types of optical fibers

core and the cladding Eq. (1) predicts a delay difference of 50 ns/km, corresponding to a bandwidth of 20 MHz · km. This value is too low to be of much use for high transmission capacities over long distances. Very high bandwidths can be achieved by reducing the core diameter to a value where only one optical mode is propagated. However, the small diameter core (~ 5 μm) produces problems when coupling light into the fiber and when splicing two fibers together. One way to overcome these difficulties is to compensate the different geometrical pathways in multimode fibers by equalizing the optical pathways as shown in the right-hand part of Fig. 4. Optimum compensation is possible provided the index profile as a function of the radius has a parabolic or near parabolic form[7)]. Fibers of this type are referred to as "graded index" fibers. In order to achieve fibers having a graded-index profile the preparation method should be capable of approximating the theoretically desired profile as closely as possible. Optimum profiles lead to a theoretical reduction of the delay time between the fastest and the slowest modes by a factor of $\sim 10^3$ compared to step index fibers.

Another reason for pulse broadening is material dispersion. This effect determines the lower limit of pulse broadening for a given source wavelength, spectral width and fiber material. A typical value for silica is 80 ps/(km · nm) at a wavelength of 900 nm.

2.2 Preparation Methods

Up to now three *c*hemical *v*apor *d*eposition (CVD) techniques have proved suitable for the preparation of high quality optical fibers: the "*o*utside *v*apour *p*hase *o*xidation (OVPO)" process[8)], the "*m*odified CVD (MCVD)" process[9)] and the "*p*lasma-activated CVD (PCVD)" process[10)]. The last mentioned process will be the main subject of this article. To give a better appreciation of the principles the alternative processes will be described briefly.

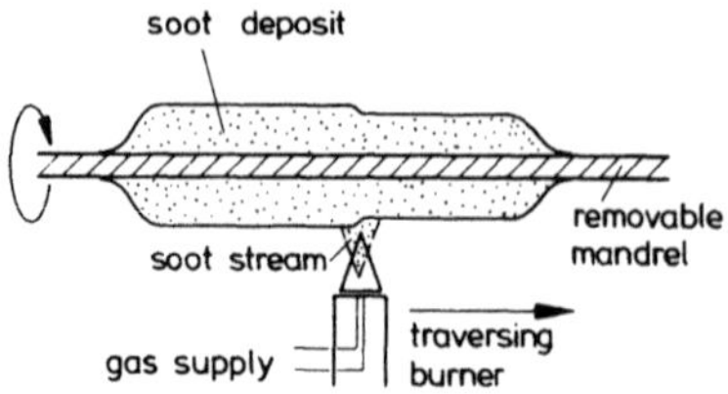

Fig. 5. Schematic view of the outside vapour phase oxidation (OVPO) process

All the processes start with highly purified and volatile compounds such as $SiCl_4$, $GeCl_4$, BCl_3, $POCl_3$, O_2, ... In the OVPO process the oxidation is carried out with the aid of a burner (Fig. 5) and the resulting soot stream of oxides is directed towards the outside surface of a rotating mandrel. During deposition the mandrel is traversed relative to the burner, which allows the deposition of uniform layers with constant or varying dopant concentration. After deposition the mandrel is removed and the porous soot preform is sintered to form a clear glass body from which the fiber is drawn.

Both step index and graded index fibers have been prepared by this process, using GeO_2 und B_2O_3 as dopant materials. Deposition efficiencies of about 45% have been obtained. The deposition rate is typically between 0.4–1.6 g/min and blanks yielding 10 km fibers have been made.

The modified CVD process (Fig. 6) has found widespread application throughout the world. The starting gases are fed into a rotating silica tube. On their way, a hot zone must be passed where the reactions to the oxides take place. The high temperature zone is produced by a moving burner heating the tube locally to a temperature of 1400–1600 °C. In common with the soot process the reaction leads to the formation of soot particles in the gas phase which, after deposition on the tube walls, are refined to a transparent layer. After deposition the tube is collapsed to a rod and a fiber is drawn (Fig. 7). By multilayer deposition the desired refractive index profiles can be approximated. Efficiencies have been reported of about 50%, depending on the compounds. Typical deposition rates are 0.1–0.37 g/min.

For both the OVPO process and the MCVD process the correct temperature setting during the course of the respective process steps has an important bearing on the quality of the final preform. The reason is that both processes consist of a first step where reactions in the gas phase lead to soot particles which settle out afterwards, and of a second step where the soot is melted to a transparent glass. In the first step the efficiency for oxide formation critically depends on the temperature of the gas and differs from one oxide to another[11]. The second step is sensitive to temperature

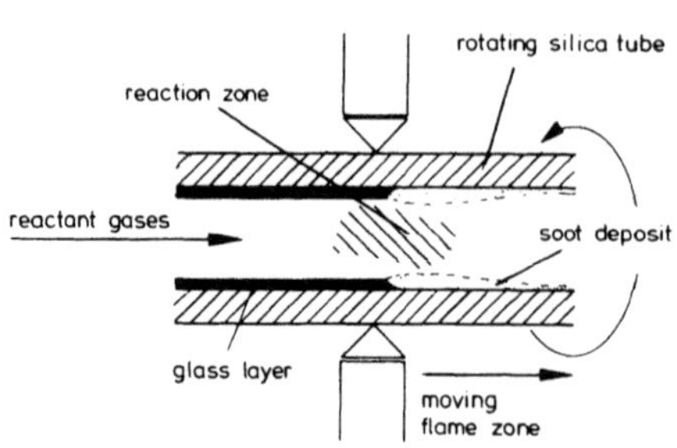

Fig. 6. Schematic view of the modified chemical vapour deposition (MCVD) process

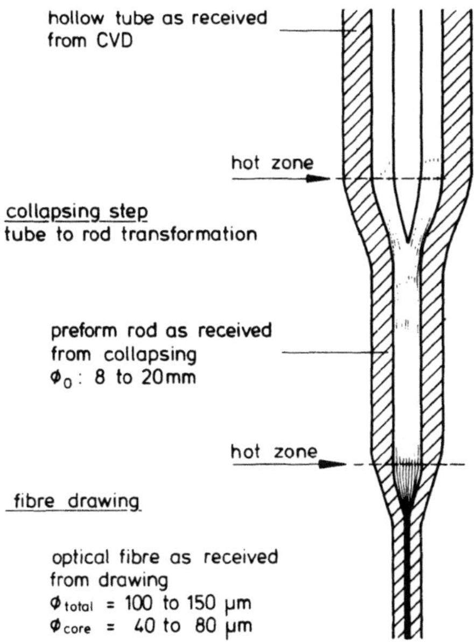

Fig. 7. Collapsing and fiber drawing stages for optical fiber preparation
Schematic drawing; in reality the three stages follow one another in time

because insufficient melting leads to bubbles which give rise to additional light attenuation in the fiber.

Therefore a plasma-activated process seemed to be attractive, which would allow the temperature to be kept as low as possible to avoid thermal reactions in the gas phase. As a consequence soot particles would also be avoided and a melting process would no longer be necessary.

3 Plasma-Activated Deposition for Optical Preform Preparation

This section gives a description of the plasma process as far as it has been described in recent publications[10, 12–14]. The individual preparation steps: coating of a silica tube on the inner surface, collapsing and drawing, are common to the MCVD and the plasma processes. However, the processes differ in the way the deposition is undertaken. This step is the most important step in fiber preparation since it determines the fiber quality to a large extent. Collapsing and drawing are important for dimensional and mechanical properties of the fiber. In the following only the deposition process by the plasma will be considered.

3.1 General Aspects

The first mention of a plasma for low-temperature deposition of glassy oxides seems to have been made by Sterling and Swann[15], who reported the deposition of amorphous silicon dioxide layers with the aid of a 1 MHz radio frequency discharge at

$1.3 \cdot 10^{-4}$ bar gas pressure in a mixture of silane and nitrous oxide or carbon dioxide. The authors found growth rates of 2–10 μm/h with the substrate temperature up to 200 °C. According to the authors this temperature is sufficient to remove water which tended to be incorporated into the films. Secrist and Mackenzie[16)] reported the deposition of silica, germania, boron oxide, titanium oxide and tin oxide layers by decomposition of various metal-organic compounds via a microwave oxygen discharge. The pressure setting was at $3.2 \cdot 10^{-4}$ bar and the substrate temperature was at some hundreds of °C. In an extensive study of the deposition of silica films by glow discharge[17)] the same authors reported a deposition rate of about $7 \cdot 10^{-4}$ mg/(cm^2 · min) of silica onto a platinum foil kept at a temperature of 306 °C.

From these papers it appeared then that low-temperature deposition of silica and other oxides which may be used as dopent materials for the preparation of optical waveguides should be possible in principle. Hydrogen-containing compounds as the starting substances can be ruled out, however, since it is known from literature[18)] that water incorporation in the layers leads to absorption peaks at 1370 nm, at 950 nm, and at 725 nm wavelength which are the first, second and third overtones of the 2730 nm OH fundamental vibration. Therefore water impurity should be as low as possible to get low attenuation in the interesting wavelength region of 0.6–1.6 μm. This last requirement led to the use of chlorides as the starting compounds.

For a fiber with a length of 1 km and with a core of 50 μm diameter, a total of about 5 g of oxide material is needed. In order to be economical the process should be capable of a deposition rate of 0.1–1.0 g/min at least. It is clear therefore that the rate had to be some orders of magnitude higher than the rates used by earlier investigators in order to fulfill the economic requirements.

There are a large number of methods for producing a plasma. Because of contamination problems in optical fiber preparation an electrodeless high-frequency plasma was selected. Initial experiments with capacitively and inductively coupled plasmas at some MHz gave unsatisfactory results because in the first case the visual appearance of the deposit was affected by the position of the induction coil. These problems were overcome by using a frequency of 2.45 GHz and by adopting a microwave cavity as the reactor. This had the further advantage that the stray fields could be much more easily screened in this ultra high frequency range.

3.2 Experimental Set-Up

The experimental set-up normally used is shown schematically in Fig. 8. The gases, various chlorides and oxygen, are supplied with the aid of flow controllers. Typical values of the gas flows Q_i are $30 \leqslant Q_{O_2} \leqslant 500$ sccm, $2 \leqslant Q_{SiCl_4} \leqslant 140$ sccm, and some tens of sccm for the other chlorides in use (sccm = standard (STP) cubic centimeter per minute). The reactor consists of a microwave cavity and a furnace capable of heating the substrate between room temperature and 1200 °C. The cavity is connected to a 2.45 GHz generator with a maximum power of 200 W. A sorption pump is used to maintain a clean atmosphere within the tube during deposition. The pressure in a typical run is selected between $1.3 \cdot 10^{-3}$ and $2.7 \cdot 10^{-2}$ bar.

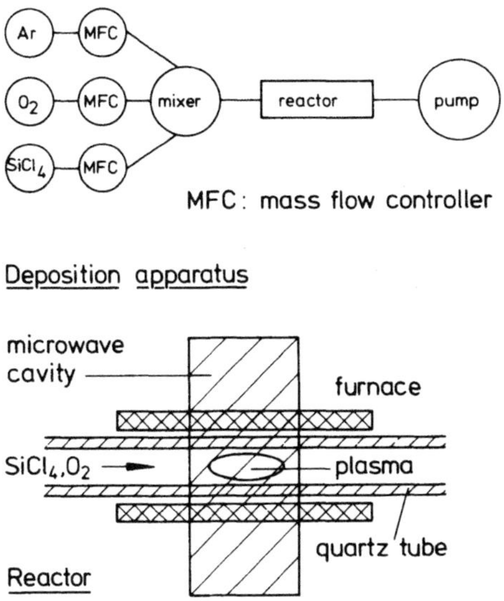

Fig. 8. Experimental set-up for the plasma-activated chemical vapour deposition (PCVD) process

The silica tube, which is normally of Heralux quality[3], is etched with dilute hydrofluoric acid on the inside, rinsed with distilled water and dried in a stream of flowing nitrogen. After installation of the tube in the reactor the tube is evacuated to a pressure below 10^{-5} bar. During this and the following steps the temperature of the tube remains at 1100 °C. In the procedure that follows, the tube is kept under a constant flow of oxygen of some tens of sccm for half an hour without changing the temperature. During this step the pressure increases into the 10^{-3} bar region. Finally an oxygen plasma is traversed over the deposition area for 10 min.

All gases used are of high quality. Nitrogen is of 99.999% and oxygen of 99.998% purity. Silicon tetrachloride[4], germanium tetrachloride[5] and boron trichloride[6] are of electronic quality.

3.3 Local Deposition of Silica[12)]

In the experiments reported in this section the cavity is stationary and the time dependence of the local deposition is followed. The mass flows Q_i, the pressure p inside the tube, the power P_m and the substrate temperature T_s are held constant.

3.3.1 Deposition Profile of Silica: Experimental Results

Figure 9 shows typical results selected for three different periods of deposition out of a series of experiments between 5 and 35 min with 5 min as the increments. The

3 Heralux: tradename of Heraeus quartz glass articles.
4 Silicon tetrachloride from Wacker Chemie.
5 Germanium tetrachloride from Preussag.
6 Boron trichloride from Elektroschmelzwerk Kempten.

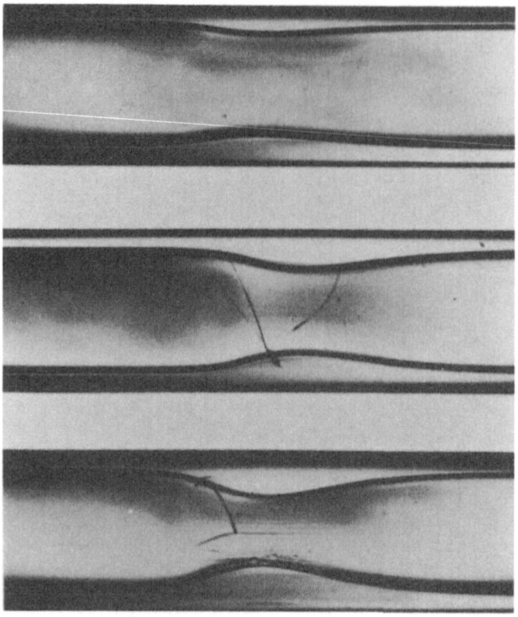

Fig. 9. Static deposition of SiO_2 [12]. T_s = 970 °C, Q_{O_2} = 30 sccm, Q_{SiCl_4} = 3.2 sccm, p = $1.1 \cdot 10^{-2}$ bar, p_m = 75 watts, initial tube i.d. = 6 mm. The flow of the gases is from left to right. (Reproduced by permission of The Electrochemical Society)

conditions are specified in the caption. The photographs were obtained from tubes which had been inserted into a liquid whose refractive index was slightly different from that of pure silica after the deposition experiment. The focus of the microscope was at the tube axis. Note that some cracks can be distinguished in the deposition area. These cracks appeared on cooling the substrate down to room temperature. They can be avoided if the substrate temperature is increased beyond 1100 °C. Channels are visible at the neck of the tube which are more prominent as the cross section becomes smaller. At the small cross sections hemispherical defects can be seen on top of the layer. These are not observed in optical waveguide fabrication where the change of the inner cross section of the tube is always small. In the following it will be assumed that the deposit is perfectly smooth.

From Fig. 9 it can be seen that the deposition profile is steeper on the side of the incoming gas flow. Further it will be noticed that the deposit possesses rotational symmetry. Since the tube is not rotated during deposition it is obvious that the influence of the gravitational force can be ignored in the explanation of the profile form. When the location of the maximum is compared with the middle of the microwave cavity it can be seen that it is shifted in the direction of the incoming gas flow. This shift strongly depends on the pressure in the tube and on the microwave energy applied, the maximum more displaced the lower the pressure and the greater the energy.

To obtain more quantitative insight the logarithm of $m_1/R^2 \pi \rho$ is plotted in Fig. 10 against x, the length coordinate of the tube. Here m_1 is the deposited mass

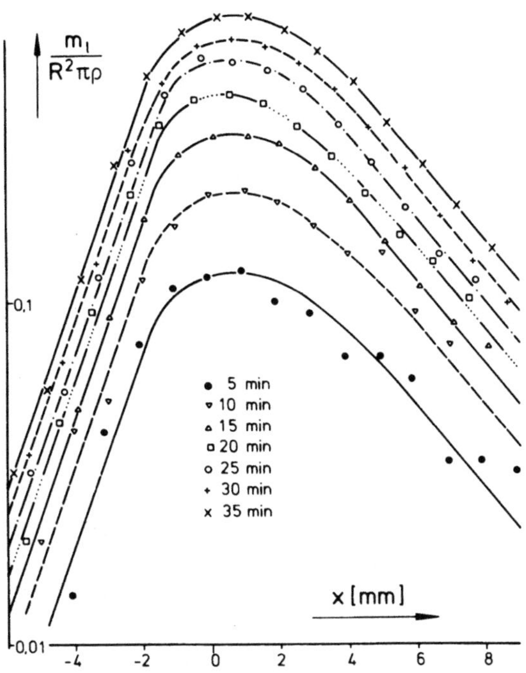

Fig. 10. Deposited mass of silica per unit length m_l as a function of the length coordinate of the tube x[12]. The flow of the gases is from left to right. (Reproduced by permission of The Electrochemical Society)

per unit length, R is the inner radius of the tube and ρ is the density of silica. The zero of x is chosen arbitrarily. By definition $0 \leqslant m_l/R^2 \pi \rho \leqslant 1$. As can be seen, the increase and the decrease are both exponential. It is clear from the Figure that the curves can be brought to coincidence by shifting them vertically. With respect to the logarithmic plot this means that

$$m_l(x, t) = g(t) \cdot m_l^0(x). \tag{2}$$

If, with fixed x, m_l is plotted for different times from 5 min to 35 min, g(t) is found to be a linear function. Thus

$$m_l(x, t) = m_l^0(x) \cdot t. \tag{3}$$

This result is trivial for short times where the thickness of the deposited layer is thin compared with the diameter of the tube. But in our case it is valid to a value where the diameter of the tube is halved – or, which is of more physical importance since it determines the gas flow velocity – where the cross-sectional area is only one fourth of the initial value.

3.3.2 Deposition Profile: Discussion

Since no investigations concerning the underlying basic steps of the deposition have been made, a simple phenomenological model will be considered and the results will be compared with the observations. Before doing so we mention the fact that the efficiency of the plasma process is complete within an experimental error of some per cent. This means that all silicon tetrachloride entering the tube is oxidized to silica and deposited upon the walls. This result has been confirmed by moving the reactor repeatedly at constant velocity over some length of the tube and by finally measuring the thickness of the deposit. When the value of the constant flow of silicon tetrachloride and the deposition period are known, the efficiency of the reaction can be calculated.

The high efficiency may be better understood if one recognizes that in a mixture of silicon tetrachloride and oxygen the chemical equilibrium is totally shifted to silica at the temperatures we apply. The activation energy for the reaction prevents the thermal reaction from proceeding rapidly under the experimental conditions chosen. Thus the plasma merely acts as a "catalyst" to overcome the activation energy of the reaction.

The model which is considered now assumes the optimum and simplest case: the silicon tetrachloride molecules passing the tube are activated on entering the plasma region and diffuse in all directions, ready for deposition on the tube walls. This is somewhat oversimplified but the consequences should show where the model holds. The term "activate" is used in ignorance of the true mechanisms, but for our purpose of a phenomenological description this knowledge is not necessary. Since we have laminar flow the equations that describe the deposition are

$$2v_m\left(1-\left(\frac{r}{R}\right)^2\right)\frac{\partial c}{\partial x} = D\left(\frac{\partial^2 c}{\partial r^2}+\frac{1}{r}\,\frac{\partial c}{\partial r}+\frac{\partial^2 c}{\partial x^2}\right) \tag{4}$$

and

$$j_{r=R} = -\,D\,\frac{\partial c}{\partial r}\bigg|_{r=R} \tag{5}$$

with the boundary conditions for our problem of rotational symmetry

$$c(R,x)=0, \quad c(r,o)=f(r) \tag{6}$$

where c is the concentration of the active species, r the radial coordinate, x the coordinate in length of the tube, R the inner radius of the tube and where f(r) is the concentration of the active species at $x = 0$. v_m gives the averaged velocity of the flow, D is the binary diffusion coefficient of the species within oxygen atmosphere and j_r is the radial flow density of the species to the wall. The flow density $j_{r=R}$ is directly connected to the deposition profile defined in Eq. (3):

$$m_1^0(x) = j_{r=R}(x)\cdot 2\cdot\pi\cdot R\cdot m_{SiO_2} \tag{7}$$

where m_{SiO_2} is the molecular weight of SiO_2 in grams.

The solution of Eqs. (4) – (6) can be written in the form of a series which converges rapidly. Therefore we only take the most prominent first term and obtain

$$m_1^0 = \begin{cases} A \cdot \exp\left(-\gamma^+ \cdot \dfrac{x}{R}\right) & x \geqslant 0 \\ A \cdot \exp\left(\gamma^- \cdot \dfrac{x}{R}\right) & x \leqslant 0 \end{cases} \tag{8}$$

where A is a function of r and is determined by the boundary condition (6). It follows from the exact form of γ^+ and γ^- that

$$\gamma^+ = -\frac{\alpha}{2} + \sqrt{\left(\frac{\alpha}{2}\right)^2 + (x^+)^2} \qquad \text{where } \alpha = \frac{2 \cdot v_m \cdot R}{D} \qquad \gamma^- = \frac{\alpha}{2} + \sqrt{\left(\frac{\alpha}{2}\right)^2 + (x^-)^2} \tag{9}$$

and where x^+, x^-, respectively, are determined by the boundary condition. For our rough estimate theoretical considerations suggest that

$$x^+ \approx x^- \approx 2.4, \tag{10}$$

which is the first zero of the Bessel function of zero order.

Under the experimental conditions $p = 1.1 \cdot 10^{-2}$ bar, $T_s = 970\ °C$ as used for the experiment given in Fig. 9, we get for typical diffusion constants

$$D_{O_2-O_2} = 237\ cm^2/s \tag{11}$$

$$D_{SiCl_4-O_2} = 76\ cm^2/s \tag{12}$$

$$D_{SiO_2-O_2} = 112\ cm^2/s. \tag{13}$$

Diffusion constant (11) is obtained from Landolt-Börnstein tables[19], (12) and (13) are calculated from

$$ {}^{T}_{p}D_{jk} = \frac{17.2\,(1 + \sqrt{M_j + M_k})}{p(V_j^{1/3} + V_k^{1/3})^2\,\sqrt{M_j\,M_k}} \cdot \left(\frac{T}{273.16}\right)^{n_D} \tag{14}$$

with p in atm, D in cm^2/s, and $n_D = 1.65$, where V_i is the molar critical volume and M_i is the molecular weight of species i.

From the evaluation of the experiment shown in Fig. 9 we get

$$\gamma^+_{exp} = 0.8 \qquad \gamma^-_{exp} = 1.9.$$

Inserting the value $\gamma^{+}_{exp} = 0.8$ into Eq. (9) and using $v_m = 847$ cm/s as obtained from the experimental conditions we get

$$D = 79\ cm^2/s.$$

This value is close to the binary diffusion coefficient given in Eq. (12). Having in mind the qualitative nature of our discussion the good agreement should not be overemphasized. The assumption of an abrupt transition boundary is incorrect. In reality there will be a transition region of some length. This will lead to a smoothing out of the entrance slope. This is confirmed by evaluation D from γ^{-}_{exp} which leads to an inconsistent result. If $v_m = 0$ then $\gamma^{+}_{exp} = \gamma^{-}_{exp} = \gamma$ should hold with γ independent of the diffusion constant. If $v_m \neq 0$ then we expect $\gamma^{+}_{exp} \langle \gamma$ and $\gamma^{-}_{exp} \rangle \gamma$ which turns out to fail for γ^{-}_{exp}. Thus it is thought that γ^{-}_{exp} reflects the increasing electric field strength rather than the effect of upstream diffusion.

The main conclusion that can be drawn from our discussion is that the formation of macromolecules in the gas phase can be ruled out under the experimental conditions applied. This was not obvious from the beginning in view of the results of MCVD- and OVPO-processes where dust formation is observed in the gas phase. For the PCVD-process a gradual formation of soot in the gas phase can also be observed for increasing pressures. In the following only those experimental conditions will be considered where the soot formation can be neglected.

3.3.3 Properties of the Layers

I Influence of the Tube Temperature

The deposition proceeds even at room temperature, but it has been found that the layers then have unsatisfactory properties and are by no means suited for the preparation of an optical preform. The layers have large inner stresses and they peel off if the thickness increases to values in the μm range. It is thought that the chlorine content is rather large at low temperatures and that the deposit looks more like a silicon-oxygen-chlorine polymer, similar to what has been found by other authors in the case of a SiF_4/O_2 gas mixture[20]. If the temperature of the tube is increased to some hundred degrees centigrade the layers no longer peel off but they may form cracks on cooling down to room temperature. These cracks totally disappear at deposition temperatures higher than ~ 1100 °C.

II Impurity Content of the Layers

As mentioned earlier, the starting materials are of high purity. Because we work in a closed system and because we have an electrodeless discharge there should be no sources of additional impurities. Neutron activation analysis revealed that all the transition metal impurities that strongly affect the transmission properties of the optical fibers are lower than 1 ppm. From fiber transmission measurements we know that, besides traces of OH, some impurities must be lower than 1 ppb because only the intrinsic attenuation of the material is found. The chlorine content is rather large at 0.1%, even at the deposition temperature of 1000 °C. Fortunately the chlorine does not affect the optical properties in the interesting region of 0.6 μm – 1.5 μm.

3.4 Co-Deposition of Silica and Germania[13)]

For the light guiding effect of optical fibers it is important to have a core with a refractive index somewhat larger than that of the surrounding cladding. Therefore the core normally consists of silica doped with other oxides to increase the refractive index. In the following we therefore discuss the simultaneous deposition of silica and germania in the silica tube.

3.4.1 Local Deposition of Doped Silica

As described before, the cavity is held in a fixed position. Instead of silicon tetrachloride alone, now silicon tetrachloride together with germanium tetrachloride enter the tube and react with oxygen to form the corresponding oxides. The deposition profile is given in the lower part of Fig. 11. The detailed experimental conditions are given in the caption. Since the germanium tetrachloride flow is only 5% of the silicon tetrachloride flow, it is not surprising that there is no essential difference between Fig. 10 and Fig. 11. The different slopes can be accounted for by the different experimental conditions, as is also to be expected from Eq. (9).

Microprobe analysis of the deposit parallel to the tube axis gave the concentration of germanium within the profile. The upper curve of Fig. 11 gives the result of the concentration C_{GeO_2} within the layer in arbitrary units. If both the local deposition profiles of silica ${}^{Si}m_1^0(x)$ and of germania ${}^{Ge}m_1^0(x)$ were identical in position and shape, a constant concentration would result. In fact this is not the case, as can be seen from the Figure. Having regard to our previous discussion on the deposition mechanisms for pure silica, this could hardly ever be expected. First, there is no reason why the transition boundary to the activated species should be at the same position for both oxides, and second, from Eq. (9) we expect the deposition profilc to

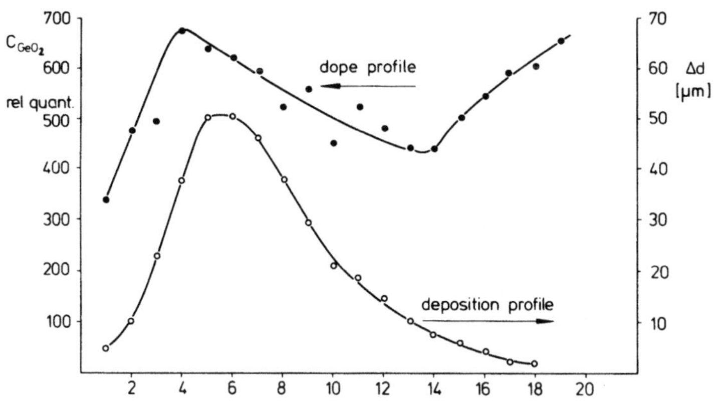

Fig. 11. Simultaneous deposition of SiO_2 and GeO_2[13)]. $T_s = 900\ °C$, $Q_{O_2} = 50$ sccm, $Q_{SiCl_4} = 2.7$ sccm, $Q_{GeCl_4} = 0.14$ sccm, $p = 8 \cdot 10^{-3}$ bar, $P_m = 65$ watts. The flow of the gases is from left to right. (Reproduced by permission of the Electrochemical Society)

have a longer tail in the case of germania because of its greater molecular weight. This is indeed observed, germania being found even beyond the area measured in Fig. 11. It is interesting to note that the actual shape of the germanium concentration shows a maximum and a minimum, but this has not been fully investigated up to now. Of course, as for the profile of silica and germania alone, the concentration profile also depends on the experimental conditions.

3.4.2 Deposition of Doped Silica Layers

As pointed out in the introductory sections, a layer of some length with constant thickness has to be deposited in order to get an optical preform. Therefore we shall now discuss the consequences of the shape of the local deposition profiles on the properties of a single layer of doped silica.

I Reactor Movement in the Direction of the Gas Flow

If the plasma, and therefore the dope profile of Fig. 11, is moved along the tube with constant velocity, a layer of constant thickness is deposited. Because the concentration of germanium varies over the length of the local deposition profile the concentration shows a variation over the thickness of the deposited layer. This is illustrated in Fig. 12a for the case where the reactor is moved with the gas flow. From the deposition profile and from the concentration curve of Fig. 11 the actual variation of the germanium concentration over the layer thickness can be calculated. For detailed information the reader is referred to the original paper[13]. The experimental result is shown in Fig. 13a. The photograph displays the cross-sectional view of a collapsed preform with only one layer deposited. The overall view gives the secondary electron picture. The areas of high germanium dope are clearly visible because the coefficient for secondary electron emission is larger than that of silica. Further, a germanium microprobe trace through the center of the preform is also given. Figure 13a agrees qualitatively very well with Fig. 12a.

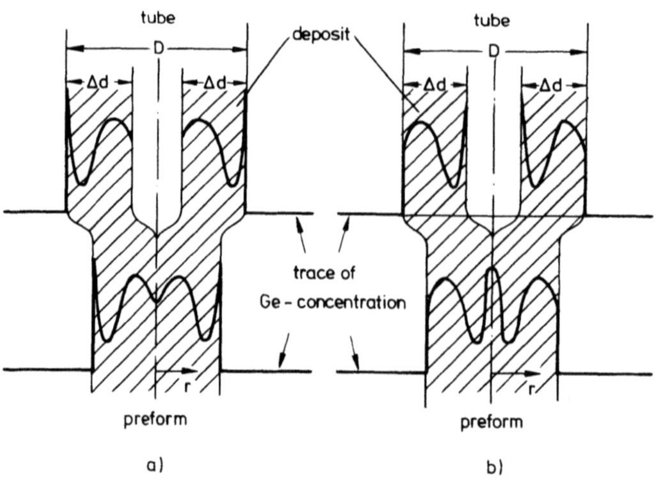

Fig. 12 a and b. Variation of dope concentration over the deposit thickness for single layer deposition (schematic)[13] **(a)** moving the reactor with the gas flow, **(b)** moving the reactor toward the direction of the gas flow. (Reproduced by permission of The Electrochemical Society)

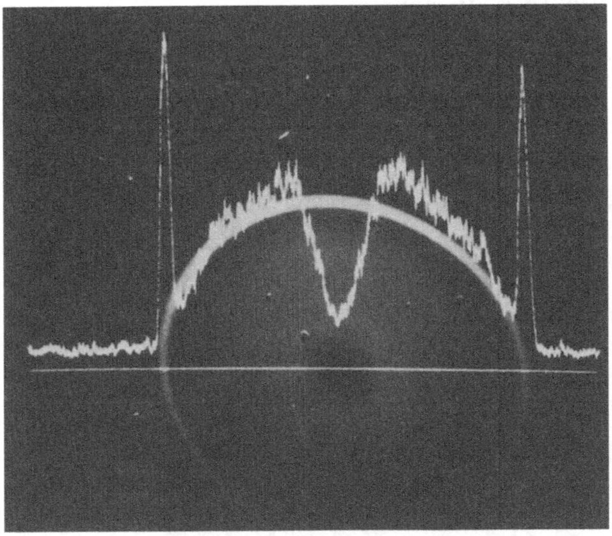

Fig. 13a. Microprobe analysis of preform core. Reactor moved in the direction of the gas flow[13].
T_s = 800 °C, Q_{Q_2} = 50 sccm,
Q_{SiCl_4} = 2.9 sccm,
Q_{GeCl_4} = 0.3 sccm,
$p = 1.1 \cdot 10^{-2}$ bar,
P_m = 150 watts.
Doped core ~ 1 mm diameter.
(Reproduced by permission of The Electrochemical Society)

II Movement of the Reactor Against the Gas Flow

This case is given in Fig. 12b and Fig. 13b. In contrast to the predictions of Fig. 12b, a deep dip occurs in the center of the concentration measurement. This is due to the depletion of germanium in the outer layers during the collapsing step because of the evaporation of germanium oxide at these high temperatures. This effect also occurs in Fig. 13a, but it is hidden somewhat by the dip already expected from Fig. 12a.

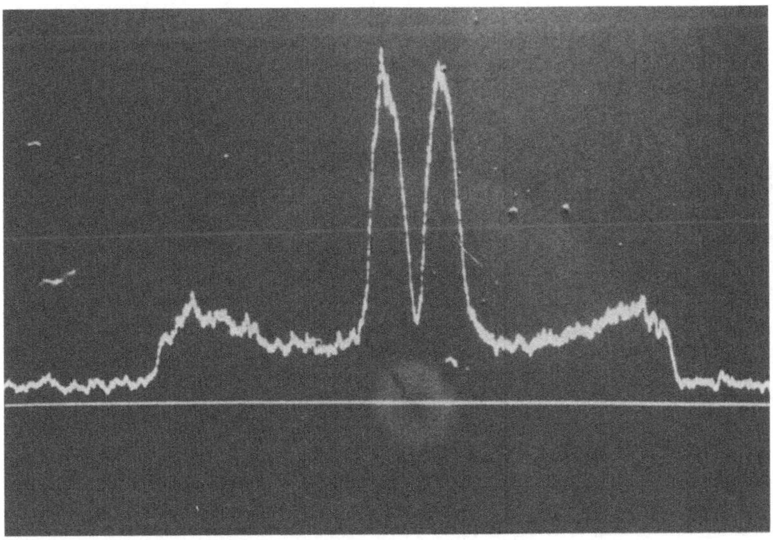

Fig. 13b. Microprobe analysis of preform core. Reactor moved against the gas flow direction[13]. T_s = 1040 °C, Q_{O_2} = 55 sccm, Q_{SiCl_4} = 2 sccm, Q_{GeCl_4} = 0.2 sccm, $p = 9.3 \cdot 10^{-3}$ bar, P_m = 90 watts. Doped core ~ 1.5 mm diameter. (Reproduced by permission of The Electrochemical Society)

3.4.3 Deposition of a Doped Multilayer Structure

In the case of multimode optical fibers a well defined radial variation of dope concentration in the fiber core is required for high data transmission capacity. Therefore

Fig. 14a. Interference micrograph of the core of a step index preform built up of 46 individual layers[13]. T_s = 900 °C, Q_{SiCl_4} = 2.8 sccm, Q_{GeCl_4} = 0.3 sccm, p = 1.3 · 10^{-2} bar, P_m = 100 watts. Doped core ~ 0.6 mm radius. (Reproduced by permission of The Electrochemical Society)

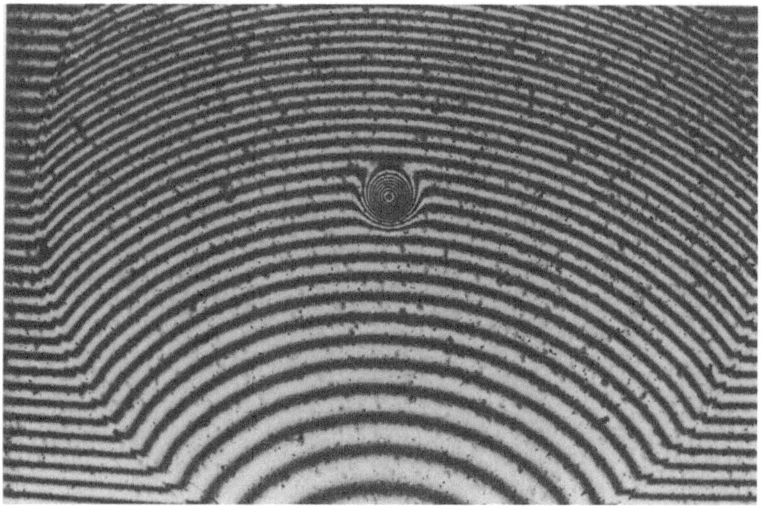

Fig. 14b. Interference micrograph of the core of a graded index preform built up of 2000 layers[13]. T_s = 1000 °C, Q_{O_2} = 50 sccm, Q_{SiCl_4} = 2.5 sccm, Q_{GeCl_4} = 0.3 sccm, p = 2 · 10^{-2} bar, P_m = 100 watts. Doped core ~ 1.0 mm radius. (Reproduced by permission of The Electrochemical Society)

it is important that the process is capable of multilayer deposition resulting in a smooth approximation of the desired index profile.

Figure 14a shows a double beam interference photomicrograph of a preform core where the preform was built up of 46 layers by moving the reactor back and forth over some length. After deposition the tube was collapsed into a rod and a slice was polished and coated with aluminium on one side. Though the flow was not changed during the experiment, the individual layers can clearly be detected as a consequence of the foregoing discussion.

Each layer has its own germanium concentration variation over its thickness, which finally results in the wave-like appearance of the interference lines. This example shows the problems that might rise if an attempt is made to approach a refractive index profile by deposition of only a few layers, even in the case of a step index profile.

Figure 14b displays a graded index profile which was produced by the deposition of 2000 layers. Because of the thinness of the layers they can no longer be detected individually. For both preforms the germanium depletion in the center is obvious from the circular interference lines.

Other arbitrary dope profiles can simply be realized by proper variation of the flow of germanium tetrachloride during deposition.

3.5 Deposition of Silica Doped with Materials Other than Germanium

Many materials are known which change the refractive index of silica on doping. For optical fibers the additional requirement is that there should be no additional losses on doping. Up to now GeO_2, P_2O_5, Al_2O_3, B_2O_3, TiO_2, and F have proved to be suitable for such purposes but there may be more. In the following we discuss the dopants that have been studied with the plasma process.

3.5.1 Deposition of Silica from $SiCl_4/SiF_4/O_2$ Gas Mixtures[14]

Previously it had been shown that plasma deposition of silica from SiF_4/O_2[20] gas mixtures at 200 °C led to silicon oxygen fluoride polymers with a refractive index lower than 1.4. Since it is also known from a recent paper[21] that fluorine-doped fibers may be used for optical communication purposes, we studied the system $SiCl_4/SiF_4/O_2$ using plasma-activated CVD. Figure 15 shows the results of the stationary experiments, with the detailed experimental conditions given in the captions. The position of the microwave cavity middle is given in the Figure. The cavity itself had an inner width of 10 mm. Figure 15a shows the result that one would have expected from the foregoing discussion: the deposition of silica from $SiCl_4/O_2$ alone is local, transparent and shifted some centimeters in the direction of the incoming gas flow. The situation drastically changes when SiF_4/O_2 is used instead of $SiCl_4$ (Fig. 15b): the deposition is expanded in area, is opaque and nearly symmetrical about the cavity position. This makes it obvious first that the simple model that was used to describe silica deposition in Sect. 3.3.2. is restricted to the system $SiCl_4/O_2$,

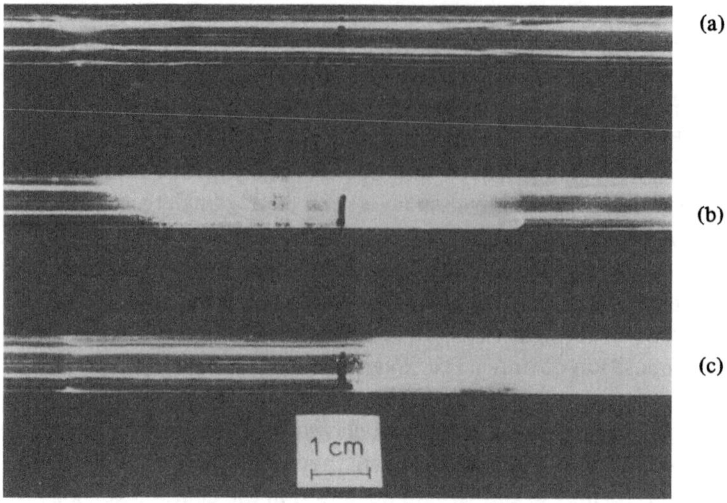

Fig. 15a–c. Stationary deposition experiments[14]. Tube diameters o.d. = 10 mm, i.d. = 8 mm, $T_s = 1100\ ^\circ C$, $p = 1.1 \cdot 10^{-2}$ bar, $P_m = 170$ watts,
(a) $Q_{SiCl_4} = 9$ sccm, $Q_{O_2} = 110$ sccm
(b) $Q_{SiF_4} = 4$ sccm, $Q_{O_2} = 110$ sccm
(c) $Q_{SiCl_4} = 9$ sccm, $Q_{SiF_4} = 4$ sccm, $Q_{O_2} = 110$ sccm.
The gas flow is from left to right. (Reproduced by permission of The Electrochemical Society)

and second that the silicon-transporting compound itself may have a drastic influence on the deposition kinetics. Figure 15c gives the result of simultaneous deposition from $SiCl_4/O_2$ and SiF_4/O_2. Without further discussion it is mentioned that this figure is not just a superposition of the two previous examples.

The change of the refractive index with changing ratio Q_{SiF_4}/Q_{SiCl_4} while the flows $Q_{SiX_4} = Q_{SiF_4} + Q_{SiCl_4}$ and Q_{O_2} remained constant is given in Fig. 16. The

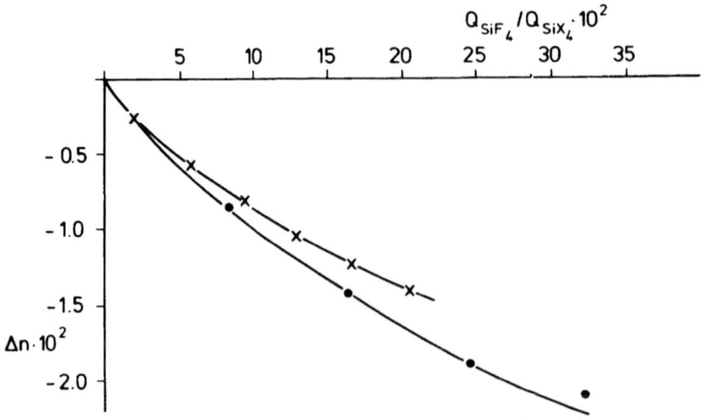

Fig. 16. Change of refractive index as a function of silicon tetrafluoride in the gas phase[14]. $Q_{SiX_4} = 10$ sccm. (Reproduced by permission of The Electrochemical Society)

Fig. 17. Cross section of a fluorine doped graded index fiber etched with dilute HF[14]. Graded core of ~ 35 μm diameter. (Reproduced by permission of The Electrochemical Society)

crosses and the dots belong to two different microwave cavities. Recently it has been reported that the incorporation of up to 3 at% of fluorine in silica decreased the refractive index linearly[22]. A decrease of $\Delta n = 0.5\ 10^{-2}$ per at% was found. From Fig. 16 thus we expect a maximum of 4 at% of fluorine in our layers. We have also found that the fluorine content increases with decreasing substrate temperature.

From Fig. 15 one would be sceptical of successfully preparing fluorine-doped optical fibers. However, experiments on deposition of fluorine-containing layers by moving the reactor over some length resulted in an excellent deposit which was absolutely clear over the deposition length. Figure 17 shows an etch figure of a fluorine-doped fiber with grading of the refractive index in the core. First 400 layers of pure silica were deposited, followed by 1070 layers with constant fluorine concentration and by 820 layers with a decreasing fluorine content. The fluorine-doped areas are clearly visible, with the etch rate being greater for larger fluorine content. The etch Figure directly resembles the form of the refractive index profile: in the ring shaped cladding the refractive index is lowered compared to the surrounding substrate and in the core the index gradually increases from the value in the cladding to the substrate value.

3.5.2 Incorporation of Boron Oxide

Boron oxide can easily be deposited from BCl_3/O_2 gas mixtures. It is normally incorporated in the first buffer layer because it decreases the softening point of silica and therefore replaces the fire-polishing step often applied to guarantee a starting tube with smooth surfaces.

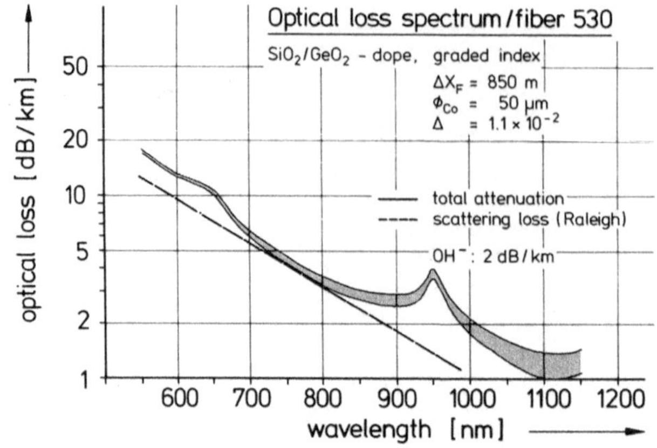

Fig. 18. Typical optical attenuation curve of a fiber produced by means of PCVD. The upper and lower curve, respectively, indicate the accuracy of the measurement

Besides the binary systems discussed so far, the tertiary system $SiO_2/GeO_2/B_2O_3$ has also been deposited from a $SiCl_4$–$GeCl_4$–BCl_3–O_2 gas composition, giving excellent optical preforms.

3.6 Optical Properties of the Resulting Fibers

The optical fibers resulting from the plasma deposition process show no particularly different properties as regards attenuation between 600 nm and 1100 nm compared with those of fibers prepared by thermal means. Figure 18 shows a typical attenuation curve with the losses near the intrinsic limit. The water peak at about 950 nm corresponds to an impurity level of ~1 ppm OH.

Since the reaction zone in the plasma process is capable of high speed, a large number of layers can be deposited per unit time. This allows smooth approximation of a desired index profile. Figure 19 shows the impulse response of a fiber[23] with

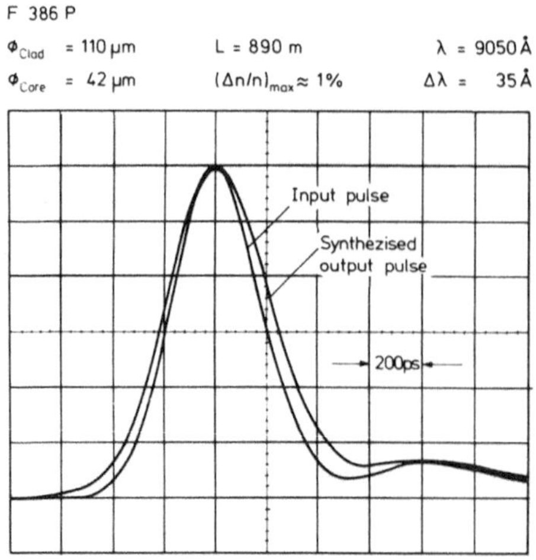

Fig. 19. Pulse broadening due to mode dispersion in a fiber with near optimum refractive index profile[23]. (Reproduced by the permission of IEE)

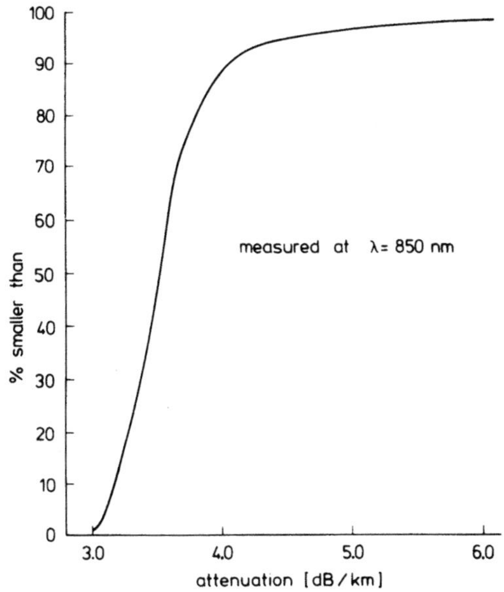

Fig. 20. Cumulative distribution of the attenuation values of 200 successive fibers[24]. (Reproduced by permission of the Instituto Internazionale delle Comunicazioni)

a refractive index profile which is near the optimum. From the evaluation it turns out that the pulse broadening from mode dispersion is lower than 150 ps/km.

The PCVD process has now been in operation under pilot plant conditions for some years. Figures 20 and 21 give the attenuation and full width half maximum pulse broadening of 200 successive fibers selected at random from the production[24]. The Figures reflect the good control that can be achieved by the PCVD process under production conditions.

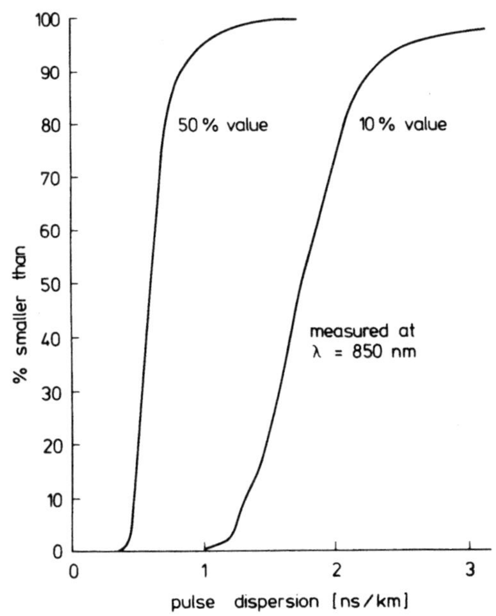

Fig. 21. Cumulative distribution of the pulse dispersion values of the same fibers as used in Fig. 20. Both the 50% and the 10% values are given which indicate the broadening at 50% and 10% of the pulse height, respectively[24]. (Reproduced by permission of the Instituto Internazionale delle Comunicazioni)

4 Summary

This chapter gives an introduction into the preparation of optical fibers by means of the plasma-activated chemical vapour deposition (PCVD) method.

First of all the physical properties of optical fibers and some of the factors influencing these properties are briefly discussed. From this it follows that preparation methods for optical fibers have to fulfill extreme requirements with respect to the impurity content of the materials involved as well as the precise optical and geometrical structure of the fiber.

Up to now three preparation methods, namely the OVPO, MCVD and PCVD method have been proven suitable for the preparation of high quality optical fibers. A short review of the OVPO and the MCVD method is given. A more detailed discussion of the PCVD method is presented. Emphasis has been laid on the description of experiments in which pure and GeO_2-oxide doped silica have been deposited. It turns out that the PCVD method has some unique properties such as deposition without soot formation in the gas phase, moderate substrate temperature, high deposition efficiency and the possibility of rapid reactor movement. Besides germanium, other dopant elements such as boron and fluorine have also been successfully deposited simultaneously with silica.

High quality optical fibers obtained using the PCVD method demonstrate that the plasma activated deposition fulfills the extreme requirements for optical fibers almost ideally. This is underlined by the optical properties of a large number of fibers prepared under the same PCVD conditions on a pilot plant scale which show a narrow distribution in their attenuation and pulse broadening values. This in turn is directly correlated with the good process control which is achievable.

5 References

1. Kao, K. C., Hockham, G. A.: Proc. Inst. Electron. Eng. *113,* 1151 (1966)
2. Jones, M. W., Kao, K. C.: J. Sci. Instrum. *2,* 331 (1969)
3. Kapron, F. P., Keck, D. B., Maurer, R. D.: Appl. Phys. Lett. *17,* 423 (1970)
4. Schultz, P. C.: J. Am. Ceram. Soc. *57,* 309 (1974)
5. Gossink, R. G.: J. Non-Cryst. Solids *25,* 112 (1977)
6. Osanai, H. et al.: Electron. Lett. *12,* 549 (1976)
7. Olshansky, R., Keck, D. B.: Appl. Optics *15,* 483 (1976)
8. Keck, D. B., Schultz, P. C., Zimar, F.: U.S. pat. 3,737,292, 1973
9. McChesney, J. B. et al.: X Int. Cong. Glass, Kyoto 1974
10. Küppers, D., Lydtin, H., Rehder, L.: Auslegeschrift 2444100 (1974)
11. Hammond, C. R., Norman, S. R.: Opt. Quant. Electr. *9,* 399 (1977)
12. Koenings, J., et al.: Chemical Vapor Deposition, Fifth International Conference, J. M. Blocher, Jr., H. E. Hintermann, and L. H. Hall, (eds.). The Electrochem. Soc. Softbound Symposium Series, Princeton 1975
13. Küppers, D., Koenings, J., Wilson, H.: J. Electrochem. Soc. *123,* 1079 (1976)
14. Küppers, D., Koenings, J., Wilson, H.: J. Electrochem. Soc. *125,* 1298 (1978)
15. Sterling, H. F., Swann, R. C. G.: Phys. Chem. Glasses *6,* 109 (1965)
16. Secrist, D. R., Mackenzie, J. D.: Chemical Reactions in Electrical Discharges (Adv. Chem. Ser., no 80), Amer. Chem. Soc., Washington, D. C., 1969, pp. 242–249

17. Secrist, D. R., Mackenzie, J. D.: J. Electrochem. Soc. *113,* 914 (1966)
18. Keck, D. B., Maurer, R. D., Schultz, P. C.: Appl. Phys. Lett. *22,* 307 (1973)
19. Diffusion in Gasen, Landolt-Börnstein, 6. Aufl., Bd. II/5a, p. 516
20. Secrist, D. R., Mackenzie, J. D.: Polymer Letters *4,* 537 (1966)
21. Mühlich, A., et al.: First European Conf. on Optical Fiber Communication, IEE, London, Sept. 1975
22. Rau, K., Mühlich, A., Treber, N.: Digest of Optical Fiber Transmission II, Williamsburg, 1977 IEEE
23. Hazan, J. P., Bernard, J. J., Küppers, D.: Electron. Lett. *13,* 540 (1977)
24. Peelen, J. G. J., Versluis, J. W., Verwaart, A. P.: Fourth European Conf. Optical Communication, Genova, 1978, Suppl. Conf. Proc., p. 69

Received April 9, 1979

Subject Index

Author Index Volumes 26–89

The volume numbers are printed in italics

GPSR Compliance

The European Union's (EU) General Product Safety Regulation (GPSR) is a set of rules that requires consumer products to be safe and our obligations to ensure this.

If you have any concerns about our products, you can contact us on ProductSafety@springernature.com

In case Publisher is established outside the EU, the EU authorized representative is:

Springer Nature Customer Service Center GmbH
Europaplatz 3
69115 Heidelberg, Germany

Batch number: 10620881

Printed by Printforce, the Netherlands